HOUSING, THE ENVIRONMENT AND OUR CHANGING CLIMATE

Edited by Christoph Sinn and John Perry

Foreword by Tony Juniper of Friends of the Earth

Chartered Institute of Housing Policy and Practice Series

in collaboration

with the Housing Studies Association

The Chartered Institute of Housing

The Chartered Institute of Housing (CIH) is the professional body for people involved in housing and communities. We are a registered charity and not-for-profit organisation. We have a diverse and growing membership of over 21,000 people – both in the public and private sectors – living and working in over 20 countries on five continents across the world. We exist to maximise the contribution that housing professionals make to the wellbeing of communities. Our vision is to be the first point of contact for – and the credible voice of – anyone involved or interested in housing.

Chartered Institute of Housing
Octavia House, Westwood Way, Coventry, CV4 8JP
Tel: 024 7685 1700
Email: customer.services@cih.org
Website: www.cih.org

The Housing Studies Association

The Housing Studies Association promotes the study of housing by bringing together housing researchers with others interested in housing research in the housing policy and practitioner communities. It acts as a voice for housing research by organising conferences and seminars, lobbying government and other agencies and providing services to members.

The CIH Housing Policy and Practice Series is published in collaboration with the Housing Studies Association and aims to provide important and valuable material and insights for housing managers, staff, students, trainers and policy-makers. Books in the series are designed to promote debate, but the contents do not necessarily reflect the views of the CIH or the HSA. The Editorial Team for the series is: General Editors: Dr. Peter Williams, John Perry and Professor Peter Malpass, and Production Editor: Alan Dearling.

Cover photographs: BRE, Peabody Trust, istockphoto (Mark Evans and Peter Garbet)
ISBN 978 1 905018 60 4

Housing, the Environment and Our Changing Climate
Edited by Christoph Sinn and John Perry
Published by the Chartered Institute of Housing
© 2008 The editors and the individual authors of the chapters

Printed on 9 Lives recycled paper by Alden Group, Witney, Oxon

*'No one could make a greater mistake than he who did nothing
because he could do only a little'*

Edmund Burke
Irish orator, philosopher, & politician (1729 – 1797)

Contents

Foreword   Tony Juniper, Director, Friends of the Earth   vii

Contributors   ix

Chapter 1:   Introduction – the changing climate   1
             John Perry and Christoph Sinn

Chapter 2:   Climate change – what is it and what causes it?   5
             John Lanchester

Chapter 3:   People and place – towards a culture of sustainability   21
             Bruce McVean

Chapter 4:   Taking the lead – local authorities and the climate
             change agenda   29
             Joanne Wade

Chapter 5:   Planning and climate change   45
             Naomi Luhde-Thompson and Hugh Ellis

Chapter 6:   Fuel poverty – social issues and sustainability   63
             Peter Lehmann

Chapter 7:   Evolution of policy and practice in low-energy housing   77
             Heather Lovell

Chapter 8:   Building new houses, are we moving towards zero-carbon?   99
             Melissa Taylor and Tom Woolley

Chapter 9:   Low impact methods of construction – the way forward?   121
             Tom Woolley

Chapter 10: Climate change and the existing housing stock   135
             Gavin Killip

Chapter 11: Social housing – 'leading' the way?   155
             Christoph Sinn, John Perry, and Adrian Moran

Chapter 12: Case study: A changing environment – creating sustainable
             housing for the future – Places for People   161
             Nicholas Doyle

Chapter 13:   *Case study:* **Black Country Housing Group**   **169**
Richard Baines

Chapter 14:   **'Greening' the housing market – putting the green**   **179**
housing market in context
Adrian Wyatt

Chapter 15:   **'Greening' your organisation**   **199**
Caleb Klaces and Adam Broadway

      Case study: The green route planner – Gentoo Group   212
Sally Hancox

Chapter 16:   **Lessons from other European countries**   **217**
Minna Sunikka

Chapter 17:   **The debate on energy and climate change – a different**
perspective   **229**
John Perry

Chapter 18:   **Conclusion – the need for action now**   **245**
Christoph Sinn and John Perry

Foreword

The debate about climate change recently reached a turning point. No longer is the discussion about whether we have a problem, or if indeed we must take urgent action to address it. The question now is instead concerned with how much we need to reduce emissions, by when and how we will do it. This shift of emphasis has not come a moment too soon.

The latest scientific assessments paint an increasingly alarming picture as to the likely consequences for people, the economy and the natural environment arising from emissions continuing at anything like business-as-usual levels. If we carry on with present patterns of energy consumption and land use, then very serious impacts can be expected to fall on human societies later this century. It is not too late to avoid the worst effects of human-induced climate change, but time is now very short indeed.

A critical threshold that is now widely accepted as a policy goal is to limit the average level of human-induced warming to below two degrees centigrade compared with pre-industrial (late 18th century) average temperature. Two degrees may not sound like a lot, but as a change in global average conditions, it is huge. To stand a reasonable chance of keeping within this level of warming, worldwide emissions of greenhouse gases must peak within a decade and then steeply decline thereafter, and to way below current levels.

Industrialised countries like the UK need to make cuts immediately. Our historical emissions are huge and with relatively strong economies natural justice demands that the rich countries begin the process of carbon reduction right away. And we should anticipate progressive reductions until we reach cuts of at least 80 per cent by 2050 compared to 1990 levels.

This is no small task, not least because alongside a real sense of urgency is considerable complexity. Unlike some other environmental challenges, climate change solutions do not exist in the form of one particular technology or one policy arena. The challenge is economy-wide and touches almost every aspect of how we live. How we travel, how much power we use, what we eat and how we dispose of waste all shape overall emissions levels. Our housing also has a big impact. Indeed, some 27 per cent of all UK emissions of carbon dioxide are linked to housing, mostly due to space and water heating and the use of electrical power for lighting and appliances.

While some of the changes we need to make in shifting to a low-carbon society are often portrayed as negative, many of the things we need to do can actually result in really positive gains: and not only for the environment, but for people and the economy as well. The housing sector is no exception. As a major part of the problem, it is also a source of many positive solutions.

For example, setting standards for new homes and a programme of incentives and support for the upgrade of existing dwellings could drive the expansion of markets of small-scale renewable energy and energy-efficiency products and technologies. The new demand could generate tens of thousands of new jobs and simultaneously deliver social benefits in the form of reduced energy bills. New national programmes that aim to reduce emissions could be deliberately tailored to maximise social advantages by being focused on those in fuel poverty, for example. If carefully focused, it is thus possible to cut emissions, improve the quality of life among the least well-off and create jobs. At a time of rising fuel prices and concerns about long-term security of energy supply, there is even more reason to make investments in houses that use less energy.

The signals sent by official policies are clearly vital in order to bring about the low-carbon transformation that we need to avoid the worst effects of human-induced climate change. Much of my work in recent years has been directed at creating national and international frameworks that will ensure carbon emissions reductions. At the same time my colleagues and I have been acutely aware of the need to secure implementation locally. After all, it is in the local context that many practical decisions are taken and it is in communities and localities that most of us live. And at the local level much can be done and in some places is already being done.

Local government has influence over many aspects of how we live and can be a hugely positive force in cutting greenhouse gas emissions. This can be done, for example, through making special planning decisions that are geared up to cutting carbon and in relation to many aspects of housing. There are already examples of leadership and these are inspiring others to see great possibilities for the creation of genuinely sustainable communities.

Uniting local action with truly global issues is not an easy task, but it is vital if we are to deal with dangerous climate change. We can create low-carbon, inclusive, inspiring communities which deliver us a better quality of life. By being positive about where we can go, in seeing the social and economic gains from taking action to avoid environmental threats, we can build places that our children will thank us for. This book sets out the thinking that can get us there. Read on, and let's work together to tackle climate change closer to home.

Tony Juniper (Director, Friends of the Earth)

Contributors

Richard Baines trained as an environmental scientist in architecture. He is currently employed by Black Country Housing Group as an Environmental Consultant and is helping the group, other housing associations and local authorities to learn about sustainability in housing and manufacturing. He is also leading research initiatives in procurement, Best Value, alternative technology and environmental business development.

Adam Broadway is an Associate Director at sustainability consultancy, Beyond Green. Adam has over 20 years of management and practical experience of the design, procurement and delivery of affordable housing. He is responsible for development management work within Beyond Green, providing advice to clients on project management, procurement and delivery options.

Nicholas Doyle is Head of Sustainable Development at Places for People, one of the UK's largest housing and regeneration organisations. He has worked in housing for 17 years and has been responsible for a wide range of energy and environmental initiatives nationally and internationally. He continues to work on a range of innovative projects and research in sustainable development.

Dr Hugh Ellis is Planning Adviser to Friends of the Earth. Hugh was a lecturer and researcher at the University of Sheffield with a particular interest in minerals, waste and public participation in planning. He has been adviser to ODPM on public participation and is a member of the TCPA policy council.

Sally Hancox is Director of Gentoo Green, the environmental sustainability arm of the Gentoo Group. She has worked in the housing industry for over 20 years rising from housing assistant to director. In addition she has spent time lecturing, providing training and working as a consultant. Her current passion is helping to deliver the sustainability agenda by harnessing the positive energy of people.

Gavin Killip is a Senior Researcher at Oxford University's Environmental Change Institute. He co-authored *40% House* in 2005 and a consultancy study for the Royal Commission on Environmental Pollution's report *The Urban Environment*, published in 2007. Gavin has ten years' experience of energy efficiency and renewable energy projects in the voluntary and public sectors. His current research focuses on delivery mechanisms for a low-carbon housing stock, particularly the role of small and medium-sized building firms. He has refurbished his own terraced house in Oxford to achieve a 65 per cent reduction in CO_2 emissions, as well as reducing other environmental impacts.

Caleb Klaces is a Consultant at sustainability consultancy Beyond Green. He works with organisations in a range of sectors to help them integrate sustainability principles into what they do, with a particular focus on internal and external communications. He is also a freelance writer and editor, with a background in literary study.

John Lanchester is a writer and former editor at the publishers, Penguin, a member of the editorial board of the *London Review of Books*, and a regular contributor to several newspapers and magazines, including *Granta*, the *New Yorker* and *The Observer*, for whom he was restaurant critic, and the *Daily Telegraph*, for whom he writes a weekly column. His first novel, *The Debt to Pleasure* (1996) won the Whitbread First Novel Award, the Betty Trask Prize, the Hawthornden Prize, and an American prize, the Julia Child Award for 'literary food writing'. The novel has been translated into 20 languages. His 2002 novel, *Fragrant Harbour*, was shortlisted for the James Tait Black Memorial Prize (for fiction). His most recent work is a book of memoir, *Family Romance* (2007).

Peter Lehmann was, from 2002 to 2008, Chair of the Government's Fuel Poverty Advisory Group and from 1999 to 2005 he was Chairman of the Energy Saving Trust. He is a non-executive Director of the Pensions and Disability Service in the Department of Work and Pensions, and a non-executive Director of Gaz de France. He is a Senior Advisor to Cool nrg, an energy-efficiency company, and is also Chair of Greenworks, a social enterprise, which resells and recycles office furniture. He was, until the end of 1998, the Commercial Director and a member of the board of Centrica, one of the British Gas successor companies. He worked for British Gas for 25 years in a wide range of jobs.

Dr Heather Lovell is a Lecturer in the human dimensions of environmental change at the University of Edinburgh. Her research and teaching interests centre on policy change and technology innovation in response to climate change, particularly in relation to housing. She has worked in academia since 1999, with experience also in environmental consulting and in the UK parliament. She has published on a range of housing and climate change issues for academic and policy audiences.

Naomi Luhde-Thompson is Planning Co-ordinator at Friends of the Earth, where she is involved in developing policy and advice for communities on planning and climate change, as well as campaigning nationally on these issues. She has also researched and written on sustainable development and public participation, both at Friends of the Earth and in her previous position at ICLEI – Local Governments for Sustainability.

Bruce McVean is a Principal Consultant for Beyond Green where he works to ensure principles of sustainability are integrated at the early stages of project development and master planning. Bruce was previously a Senior Policy Advisor at the Commission for Architecture and the Built Environment (CABE) with responsibility for developing and communicating CABE's policies on sustainable design and climate change. He is a visiting lecturer on sustainable communities and regeneration policy at London Metropolitan University.

Adrian Moran is a Policy Manager with the Housing Corporation, whose remit includes neighbourhoods and communities. His previous roles at the Corporation include Head of Innovation and Good Practice and Tenant Services Manager. Prior to this, he worked for housing associations, in a training scheme, and a local authority housing aid centre.

John Perry was Director of Policy at the Chartered Institute of Housing until 2003, when he became Policy Adviser. Since that date, he has been based in Nicaragua where he works voluntarily on sustainable development projects, including the one described in Chapter 17. He has published several books and reports, including (with David Garnett) the book *Housing Finance* and guides to housing services for asylum seekers and refugees, community participation and community cohesion. His house in Nicaragua is 'off grid' and electricity is provided by solar panels.

Christoph Sinn is a Policy and Professional Practice Officer at the Chartered Institute of Housing with a background in residential care. He is currently leading on the Institute's work around environmental sustainability. Prior to joining the CIH, he has worked for a private sector housing consultancy, the University of Central England and in residential care settings in Germany.

Dr Minna Sunikka is a Lecturer at the University of Cambridge Department of Architecture and a Fellow and the Director of Studies in Architecture at Churchill College. Before coming to Cambridge, she worked on Environmental Impact Assessment, sustainable urban regeneration and comparative policy analysis in Finland and in the Netherlands. Her research focuses on policies to improve the energy efficiency of the European housing stock and she has published several books and articles on the subject.

Melissa Taylor is a Senior Lecturer on the MSc course in Architecture: Advanced Environmental and Energy Studies, in the Graduate School of the Environment, based at the Centre for Alternative Technology. As well as delivering lectures in Measuring Sustainability, Assessment Methods and Cities and Communities, she co-ordinates thesis activities, and teaches consensus design and the use of test cells. She also works as Senior Sustainability Consultant for Hilson Moran Partnership, where she leads a team of Code for Sustainable Homes assessors, and delivers internal training, and external CPD seminars on the code.

Dr Joanne Wade is a Director of a sustainability consultancy specialising in work with the public sector. Joanne has 20 years experience in the energy and environmental field, and has worked on climate change issues with local government for the past 12 years. She is a member of the Council of the Energy Institute and Chair of the Eaga Partnership Charitable Trust.

Tom Woolley is an architect and writer based in Northern Ireland. He has divided his career between housing and community work, architectural practice and university teaching. He is now Professor of Architecture at the Graduate School of the Environment at the Centre for Alternative Technology in Machynlleth, Wales. He was an editor of the *Green Building Handbook* and also author of *Natural Building*. He is currently building a small timber frame house using natural materials like hemp and is involved in advising eco housing groups.

Adrian Wyatt is founder and Chief Executive Officer of Quintain Estates and Development PLC. Adrian is also founder and Chair of the Sustainable Environment Foundation, a charity which raises awareness of the impact of the built environment on climate change. He is a Chartered Surveyor and a Chartered Financial Analyst. Prior to the formation of Quintain, Adrian was a partner at Jones Lang LaSalle.

CHAPTER 1:
Introduction –
the changing climate

John Perry and Christoph Sinn

Why is the environment important to people who work in housing? As the typical housing manager walks into her office in the morning, it's unlikely that climate change will be at the front of her mind. She parks her car, turns on the lights (or maybe the cleaner left them on an hour earlier?), switches on her p.c. and begins the day's tasks. Maybe there will be some repairs reports to chase, the latest rent arrears returns to investigate, or a visit to a difficult tenant about her teenage children's behaviour.

If she thinks about the environment at all, it's more likely to be about the state of the office or problems of vandalism in the estate where she works. But, already, her activities so far that day have had a tiny but measurable impact on the environment in a much wider sense. Driving to work, switching on the office lights and having the computer switched on for the whole day all use energy. 'Using energy' almost certainly means petrol burnt in the car, or coal or gas burnt in the power stations that produce the electricity. These are fuels rich in carbon, which changes into carbon dioxide when the fuel is burnt, and as a gas it enters the atmosphere. Carbon dioxide is one of the main 'greenhouse' gases which trap heat and prevent it escaping into the night sky. As more of these greenhouse gases accumulate, the atmosphere is warming up. Eight of the warmest years on record around the world have occurred in the last ten years.

'So what?' – the housing manager might answer. 'It's getting warmer, but where's the problem? Milder winters means it is easier for people to heat their houses, warmer summers give us more opportunities to get a tan. OK, so we are messing with the environment, but so far things seem better not worse, so why worry?'

Before going back to that question let's take a further look at the housing manager's patch and what might be happening in it. Supposing she is responsible for 2,000 homes, then they might be producing not just a few grammes of carbon but, over a year, around 12,000 tonnes. In fact, some 27 per cent of the carbon produced in Britain comes from homes. How we run our houses is one of the biggest factors driving climate change.

But even this figure shows only part of the importance of the household sector. The other big producer of carbon is transport: mainly cars and lorries. There is no short-term prospect of persuading all those drivers – including our housing manager – to go to work or do their jobs without continuing to use petrol. Yes, some people might be willing to switch to public transport, or car-share, or trade in their SUV (Sport Utility Vehicle). This will help. But it won't give us the big reductions in carbon emissions that we need in the immediate future. That brings us back to homes – where there is considerable potential for change.

Where our homes are, how they are designed, what heating they use and what equipment they contain are all important factors. How much we use our cars (for example, if our homes are far from the shops or there is no public transport), how much energy is wasted through draughty windows or uninsulated roofs, and what domestic devices we have and how much they are used, are all issues about our lives at home.

This means that everyone who deals with housing – our housing manager, the improvements manager, contractors and, in the private sector, builders, estate agents and lenders – all have an enormous influence on energy consumption. Every day they deal with problems where energy use is a factor. Box 1.1 gives some examples about how we can all start to look at the problem in fresh ways.

But let's go back to the question 'why worry?' It helps if we think about the timescales we ordinarily use in the housing world. A finance director looking at a stock transfer, models the costs over 30 years. An improvements manager has to hit the Decent Homes Standard target by 2010. Many housing professionals are already used to looking ahead over long periods, knowing that action taken now will affect the chances of their hitting long-term targets.

It's the same with climate change. In terms of weather, next year might be very much like this. But looking ahead we know that UK summers will get hotter. Scientists believe that summer temperatures could increase by between 4° centigrade and 10° centigrade if carbon dioxide levels in the atmosphere double, as they are expected to over the next 100 years.[1] It may not seem a lot, but it will be sufficient to radically affect our daily lives – how much water we need and how we get it, how often we have severe storms, how much flooding there is, how often there are heat waves, how much subsidence there is affecting buildings.

Heatwaves, for example, lasting five days or more – like the ones experienced in 2005 and 2006, are likely to occur almost every year by as soon as 2010. Such heatwaves might even take place several times each summer. And scientists are concerned that the climate will not just become warmer then stabilise, but that overheating will get out of control as new forces come into play.

---

1   http://environment.guardian.co.uk/climatechange/story/0,,1863125,00.html

OK you say, but surely the government is tackling the issues and there is nothing we can do about it. Even if the first part of that were really true, it will depend on actions by individuals to carry out the changes that are needed. People who work in housing are in a particularly influential position.

This is why we have produced this book. The original book with a similar title, *Housing and the Environment*, published in 1994, had several references to climate change, but was broader in scope. For this edition, we judged the issue to be so important that it has become the main focus of the book. We have brought together writers and lobbyists, academic experts and practitioners to present as broad coverage as possible of how housing affects the quality of our environment, with energy use and climate change as the principle focus.

Many readers may still feel unconvinced about climate change, or unsure of what it will mean. For this reason we begin the book with a persuasive introductory chapter by journalist and writer John Lanchester, which sets the scene for what follows.

Chapters 3, 4 and 5 also help to set the context. Chapter 3 is about the cultural change needed in starting to face these issues. Chapter 4 looks at the overall role of local authorities and Chapter 5, the role of the planning system.

The next set of chapters looks in detail at housing issues. Chapter 6 considers fuel poverty, which must be an important influence on work to make houses more energy efficient. Chapter 7 looks at the recent evolution of policy on low energy housing, and the pioneer examples of low energy housing. Chapters 8 and 9 consider new-build and low-carbon technologies, while Chapter 10 considers the effort needed to upgrade the existing stock. Chapters 11, 12 and 13 summarise current policy and give two case study examples of what social housing landlords are actually doing to implement it.

The next two chapters widen the argument again to look at the 'greening' of the housing market (Chapter 14) and of housing organisations (Chapter 15). A case study shows how one housing body has attempted to do just that.

Two international chapters also provide a wider context to what is happening in the UK. Chapter 16 looks at other European examples, and Chapter 17 at the position (and points of comparison) with a developing country in Latin America. Our concluding chapter attempts to point the way forward at a very critical moment in the development of policy on housing, the environment and climate change.

Box 1.1: How can housing professionals start to engage with climate change? – here are some examples

When the **housing manager** considers her arrears reports, it surely crosses her mind that tenants are continually juggling with their regular bills, including what they spend on electricity and gas. What could she do to help? Fitting energy-efficient light bulbs could save each tenant around £9 per light bulb each year. Properly draught-proofing the houses could save a further £20. Would someone sponsor this work?

Improvements managers are more influential still. Their work will have been focused on the Decent Homes Standard, but this only achieves limited improvements in energy efficiency. They are still responsible for some two million council and housing association homes that could have cavity wall insulation – but don't. Government estimates that highly improved insulation of social housing properties could, by 2020, save no less than 500,000 tonnes of carbon emissions each year – but only if we start to take action now.

Contractors and developers are influential too. They have to put into bricks and mortar what is specified, perhaps by an enlightened architect. Some contractors are leading the way by making energy-efficiency proposals themselves, and showing their familiarity with the techniques and materials. Some developers are emphasising the energy efficiency of their homes. But there is still a big gap between intention and reality. For example, building regulations that are supposed to ensure that houses are less draughty and retain their heat better are not well-enforced.

Estate agents and lenders have a tremendous unused potential for influencing consumer behaviour. Importantly, they are dealing much of the time with second-hand homes, which are more likely to be energy-inefficient than new ones. Government has introduced Energy Performance Certificates (EPCs) which carry information on the energy efficiency of each home sold. From October 2008, EPCs will cover private and social rented housing.

Development managers can show the way by aiming not just for the standards required by the Housing Corporation, but much higher. For example, some houses in Sweden and elsewhere, with harsher climates, are so efficient that they need no space heating. Imagine the benefits for tenants if this could be done as well as for the targets we all need to try to meet.

Finally, at a **personal** level we can all do much more. Can we leave the car at home part of the time? Have we switched those lights off? Do we leave the computer running all night? Even by the time you arrive at the office in the morning, you have already had several opportunities to save energy. Just do it!

CHAPTER 2:
Climate change – what is it and what causes it?

John Lanchester

It is strange and striking that climate change activists have not committed any acts of terrorism. After all, terrorism is for the individual by far the modern world's most effective form of political action, and climate change is an issue about which people feel just as strongly as about, say, animal rights. This is especially noticeable when you bear in mind the ease of things like blowing up petrol stations, or vandalising SUVs (Sport Utility Vehicles). In cities, SUVs are loathed by everyone except the people who drive them; and in a city the size of London, a few dozen people could in a short space of time make the ownership of these cars effectively impossible, just by running keys down the side of them, at a cost to the owner of several thousand pounds a time. Say fifty people vandalising four cars each, every night for a month: six thousand trashed SUVs in a month and the Chelsea tractors would soon be disappearing from our streets. So why don't these things happen? Is it because the people who feel strongly about climate change are simply too nice, too educated, to do anything of the sort? (But terrorists are often highly educated.) Or is it that even the people who feel most strongly about climate change on some level can't quite bring themselves to believe in it?

I don't think I can be the only person who finds in myself a strong degree of psychological resistance to the whole subject of climate change. I just don't want to think about it. This isn't an entirely unfamiliar sensation: someone my age is likely to have spent a couple of formative decades trying not to think too much about nuclear war, a subject which offered the same combination of individual impotence and prospective planetary catastrophe. Global warming is even harder to ignore, not so much because it is increasingly omnipresent in the media but because the evidence for it is starting to be manifest in daily life. Even a city boy like me can see evidence that the world is a little warmer than it was.

Part of the problem is one of scale. Global warming is as a subject so much more important than almost anything else that it is difficult to frame or discuss. At the moment there is a global warming-related item on the news at least once a week. Today, for instance, there are two: close to home, a judge throwing out the government's phoney 'consultation' process over nuclear power, and further away, at a conference in Washington, an 'informal agreement' marking a new commitment to 'tackling climate change' and resulting in a 'non-binding' declaration which reflected 'a real change of mood'. Just what the world needs – more hot air. And then the news moves on to other things, to contaminated Anglo-Hungarian turkeys and gang shootings and potential schisms in the Anglican Church. There is a kind of falsehood built into this; at the very least, a powerful degree of denial. If global warming is as much of a threat as we have good reason to think it is, the subject can't be covered in the same way as church fêtes and county swimming championships. I suspect we're reluctant to think about it because we're worried that if we start we will have no choice but to think about nothing else. James Lovelock, in his powerful and extremely depressing book *The Revenge of Gaia*, says this:

> *'I am old enough to notice a marked similarity between attitudes over sixty years ago towards the threat of war and those now towards the threat of global heating. Most of us think that something unpleasant may soon happen, but we are as confused as we were in 1938 over what form it will take and what to do about it. Our response so far is just like that before the Second World War, an attempt to appease. The Kyoto agreement was uncannily like that of Munich, with politicians out to show that they do respond but in reality playing for time.'*

I may be wrong in speaking of a general sense of psychological resistance; perhaps I'm only talking about myself. In any case, with the whole topic so charged and so difficult, it is best to begin with the agreed facts.

The climate of our planet is not stable. The whole of recorded human history has taken place within what is, from the earth's point of view, a relatively narrow band of temperature. From a glaciological perspective, we are living in an ice age, because there is ice at the poles – which has by no means always been the case. Fifty-odd million years ago, not only was there no ice at the North Pole, the temperature there was 23° centigrade. Ten thousand years ago, at the end of the last cool snap of this current ice age – known as a 'glacial', to distinguish it from the warmer 'interglacial' of the type we are living through now – much of Northern Europe was buried beneath miles of ice. Sea levels were hundreds of feet lower than they are today, and there was a thousand-mile-wide land bridge between Russia and North America. According to some palaeo-climatologists, 700 million years ago, during a period known as the Varangian (for some reason geological ages are named after characters in Dungeons and Dragons), almost the whole planet iced over, well-nigh irrecoverably. Earth was prevented from becoming a permanently lifeless ball of ice only through processes which are not fully understood. This is known, with appropriate chillingness, as the 'snowball earth event'.

Most of the variation in the earth's climatic cycles comes from small irregularities in its orbit. These irregularities are magnified by the immensely complicated systems of the earth's climate. A crucial one of these is the greenhouse effect, without which there would be no life on our planet, since it is the greenhouse effect which prevents the sun's infra-red radiation from simply bouncing back out into space. The existence of the effect was first posited in 1859 by the Irish scientist John Tyndall, who said that without the greenhouse effect, '...*the warmth of our fields and gardens would pour itself unrequited into space, and the sun would rise upon an island held fast in the iron grip of frost*'. The Swedish chemist Svante Arrhenius added to Tyndall's work in the early 20th century by pointing out that human activity was adding to the level of carbon dioxide in the atmosphere. Because CO_2, along with other gases such as water vapour and methane, prevents radiation from escaping, it is a 'greenhouse gas', and therefore increased levels of CO_2 make the earth warmer – not that Arrhenius was especially worried about that, since he thought that the rate of increase would be low. The basic science of this was not in dispute, but the area was also not one of much scientific interest except to one or two mavericks.

One of them was a young American physicist called James Hansen, whose 1967 PhD thesis studied Venus and came to the conclusion that it was the greenhouse effect which made the planet so warm – 400° centigrade on the surface, hot enough to melt lead. A probe later the same year showed that the atmosphere of Venus was in fact 96 per cent carbon dioxide, and Hansen became fascinated by the greenhouse effect on earth. At the prompting of a geochemist and oceanographer called Charles David Keeling, the observatory of Mauna Loa on Hawaii had been collecting data on the level of CO_2 in the atmosphere since 1959. The result – the 'Keeling curve' – clearly showed that levels of atmospheric CO_2 were rising sharply. In 1979, Jimmy Carter asked the National Academy of Sciences to look into the question. The Ad Hoc Study Group of Carbon Dioxide and Climate did that, and reported that they had '...*no reason to doubt that climate change will result and no reason to believe that these changes will be negligible*'. Dating more or less from that report, a huge amount of work has been done on the science of the subject, and especially on the detailed, sophisticated and controversial computer models on which predictions about the future are based. (One of the world's leading centres for this research is our very own Hadley Centre, based near Bristol and run by the Met Office.)

It was 1988 before the issue of CO_2 and the climate was again raised to public attention. James Hansen testified before a Congressional hearing that he was '...*99 per cent*' certain '...*global warming is affecting our planet now*'. The attention brought by the hearings prompted the UN and the World Meteorological Organisation to found the Intergovernmental Panel on Climate Change, with a brief to study and report on the question of greenhouse gases and their effect on the climate. The IPCC's first report, in 1990, said that there was evidence of global warming to the tune of about 0.5° centigrade in the past century, that the cause of the warming to date could be as much natural as human, but that action was needed to prevent the build-up of greenhouse gases in the future. From that point on, the Panel's degree of confidence about the causes of global warming grew steadily. Its next report was in 1995, when it concluded that '...*the balance of evidence suggests a discernible human influence on global climate*'. By 2001 it had decided that it was *'likely'* that human activity had caused the greater part of the century's warming: 'likely' means that it put the probability at between 66 and 90 per cent. In 2006, the panel reported for the fourth time. The latest report says that the observed global warming over the last fifty years was *'very likely'* to be the result of human activity, a statement which means that they are between 90 and 95 per cent certain.[1] This means that from the scientific point of view there is no longer a debate about human-caused global warming and the only question left open is what exactly to do about it.

Laid out like this, it all seems pretty clear-cut: a story of speculation and research leading to a gradually increasing degree of certainty. But in practice the question of climate change has never been other than bitterly contested, to an extent that reflects structural flaws in three areas: the political context of science; the reporting of science in the media; and the more general relationship between science and the public. It can be argued that the question of how we got here doesn't matter, especially not when compared to the overarchingly urgent question of what to do next; but I'm not sure that's right. A maxim in the theory of problem-solving says: if you can't break it, it isn't fixed. In other words, if you don't know how something came to not-work, you can't be sure that you have reliably made it work. Since a systematic approach to climate change would involve a new relationship between scientific predictions and public policy, it seems a good idea to try and think clearly about how we got to this point.

The simplest issue has been to do with the politicisation of science. The story is clearest in the US, which leads the world in polluting the planet, and in studying the climate, and has set the pattern for the global debate on the issue. Unfortunately, the climate debate came along at a time when the Republican Party was wilfully embracing anti-scientific irrationalism. One way of telling this story – adopted by Kim Stanley Robinson in his novel *Forty Signs of Rain* – begins with the Scientists for Johnson Campaign, run by a group of eminent scientists who were worried about Barry Goldwater's apparent eagerness to wage nuclear war. Their campaign had a considerable impact, and when Richard Nixon got to the White House four years later he was convinced that scientists were a dangerously anti-Republican political lobby. Nixon shut down the Office of Science and Technology, and kicked the presidential science adviser out of the cabinet – an effective and still unreversed removal of science from the policy-making arena in the US. Still, the Republicans did not really go to the dark side over science until the second Bush presidency. The first Bush was willing to sign the UN Framework Convention on Climate Change, a document which was unanimously approved by the Senate, but his son has appeared

---

1   The *Summary for Policymakers* employs a distinctive scale for translating probabilities into ordinary language: 'Virtually certain >99 per cent probability of occurrence; Extremely likely >95 per cent; Very likely >90 per cent; Likely >66 per cent; More likely than not >50 per cent; Unlikely <33 per cent; Very unlikely <10 per cent; Extremely unlikely <5 per cent'. Some commentators have had fun at the expense of this, but it seems to me to verge on the genuinely useful.

reluctant to discuss the subject and even more reluctant to act – this being one of the most important ways his administration has acted as a wholly owned subsidiary of the oil industry. James Hansen and other scientists have reported attempts by the administration to prevent them from voicing their views; and the administration has consistently tried to weaken and tone down the language of its own agencies' reports on the subject. What makes this so bizarre are Bush's private views on energy and oil, as reflected in the various ecologically friendly decisions he has made at his own ranch in Crawford (it uses geothermal heat pumps, and has a 25,000 gallon underground cistern to collect rainwater), and in this passage from his speechwriter David Frum's book *The Right Man*:

> *'I once made the mistake of suggesting to Bush that he use the phrase cheap energy to describe the aims of his energy policy. He gave me a sharp, squinting look, as if he were trying to decide whether I was the stupidest person he'd heard from all day or only one of the top five. Cheap energy, he answered, was how we had got into this mess. Every year from the early 1970s to the mid 1990s, American cars burned less and less oil per mile travelled. Then in about 1995 that progress stopped. Why? He answered his own question: because of the gas-guzzling SUV. And what had made the SUV possible? This time I answered. 'Um, cheap energy?' He nodded at me. Dismissed.'*

More or less the only conclusion one can draw from that under-reported passage is that W. is well aware of the realities but has been knowingly acting as a stooge for the oil industry. He is not alone. It is shocking to learn from George Monbiot's book *Heat* just how systematic the oil lobby has been about spreading a smokescreen of doubt around the question of climate change. The techniques in play were learned by the tobacco lobby in the course of the fights over smoking and health. *'Doubt is our product since it is the best means of competing with the "body of fact" that exists in the minds of the general public,'* an internal memo from one tobacco company states. *'It is also the best means of establishing a controversy.'* Or, as the Republican pollster Frank Luntz put it in a memo to party activists during W.'s first midterms, *'Should the public come to believe that the scientific issues are settled, their views about global warming will change accordingly. Therefore, you need to continue to make the lack of scientific certainty a primary issue in the debate.'* Oil money and tobacco money have gone to bodies such as the Competitive Enterprise Institute, the Cato Institute, the Heritage Foundation, the Hudson Institute, the Frontiers of Freedom Institute, the Reason Foundation and the Independent Institute. Exxon, in particular, is a great one for sponsoring climate-denying websites and lobby groups.

This policy has been remarkably effective. While the peer-reviewed science on global warming is overwhelming – a 2004 survey in *Science* showed that of the 928 peer-reviewed papers on the subject, *'none of the papers disagreed with the consensus position'* – the balance in the media has been split almost 50-50 between the scientific evidence on the one hand and 'sceptics' on the other. On Monbiot's account, the BBC has woken up to the way in which it was 'fooled by these people', which is good news if it is true; but the corporation has hitherto been weak-minded about its reporting of climate change. The ideology of balance has led it to include the 'other side' of a debate which has, among scientists, only one side; a recent highlight was an appearance by former Chancellor of the Exchequer Nigel Lawson on Newsnight, arguing, or 'arguing', as follows: '...the whole science is extremely uncertain – that is well known to anybody who has studied it'.

The problem with 'balance' is partly a problem with the way science is reported. 'Balance' works, sort of, as a way of discussing politics in a two-party system. (Though it has to be said that the remorseless polarisation, whereby I say yah because you said boo, is one main reason for the decreased interest in party politics.) Since the climate debate has been polarised on left-right lines in the US, it has seemed appropriate to the media to treat it as a polarised issue, one on which there are two schools of thought, which, in respect of the science, it isn't: there is one school of thought, and a few nutters.

The way the issue is reported reflects the fact that there are people who want to believe in global warming, and wanted to do so right from the start, before the evidence had accumulated to the point where it was no longer an issue of belief. Similarly, there are plenty of people who did not want to believe in man-made global warming, and who are continuing to refuse to believe in it even though the balance of the evidence has changed. But we can't afford to be distracted from the factual position either by the people who want it to be true or the people who want it not to be, and there is an urgent requirement in the public arena for the issue to be considered now as one of plain fact.

When we come to sum up how we got to this point, there is one other factor to add to the politicisation of science and the reporting of science. It is a deeper, murkier consideration, and it bears on the way our society is in thrall to science and at the same time only partly understands it. Our material culture is based on science in a way so profound that our attitude to it approaches a kind of faith. Arthur C. Clarke said that '...any sufficiently advanced technology is indistinguishable from magic'. This is a remark beloved of SF fans, and endlessly quoted in discussions of what might happen if there were ever to be contact between humans and aliens (or time travel, etc.). Its real sting is that it is a description of the world we already inhabit. Electric light and power, and television, and computers, and fridges, not to mention cars and planes and lasers and CD players and dialysis machines and wireless networking and synthetic materials, are things we take on trust: we don't know how they work, but we're happy to benefit from using them. We may have a rough understanding of scientific method, and even a rough Bill Brysonish sense of some of the science involved, but that is about it; our attitude contains significant components of faith and trust and incomprehension, while at the same time we are grateful for the wonders modern science has brought us.

Our faith-based contentment with science has been challenged before, most particularly by the invention of nuclear weapons. But with global warming, science is bringing us catastrophic news, and is doing so, moreover, on the basis of predictions about the future which demand urgent and radical action in the present. It's not like being told about some scientific breakthrough, waiting a few years, and then having the breakthrough manifest itself in the form of a technology that gradually becomes more useful over time. The issue of global warming is the opposite of that: we are required to act on the basis of the faith in science which is one of the fundamental underpinnings of our society, but the faith has never been made quite so explicit before, and the need to act radically, urgently and expensively on the basis of scientific models is testing that faith to the full and beyond. This perhaps underlies the note of hysteria or of will-to-persuade that is so apparent in the public discussion of global warming. In general, the more urgent an issue is for the public, the harder scientists try to appear detached and elevated. Whenever there's a public health scare, whether it's justified (BSE) or not (MMR), government scientists begin using the word 'evidence' in every sentence, which is always a sign that they think that the real issue is a failure on the part of the media and public to understand scientific method. (I learned this through encounters with a senior biologist I got to know while reporting on BSE ten years ago. She would consistently refuse to answer any question I put to her about the implications of the discovery of human-form BSE, and the consequences for public policy: it was always *'...the evidence doesn't show this'*, *'...the evidence doesn't show that'* and *'...there is as yet no evidence to suggest'*. But I heard that in private she was going around begging other people, and especially people with children, to stop eating beef.) Working scientists have a very low opinion of the way science is reported, which shades without too much difficulty into a belief that the public is too stupid to understand science. In the case of global warming, these factors have come together to create a situation where the scientists involved are not just talking in a new way, one unfamiliar to both them and us, but are in effect trying to sell us something. And we the public might be undereducated, but we know not to trust entirely someone who is trying to sell us something. The impression that some scientists are consciously trying to make us more afraid is a potent aid to the sceptics.

We deeply don't want to believe this story. The fourth report of the IPCC makes it clear that we are right not to want to. Its *Summary for Policymakers* (SPM) is a strange document. The way the SPM works is that the scientists write a report, and then are put together in a room with representatives of the world's governments, and between them they agree a text that has full support, the idea being that there is nothing left that can be contested: that the SPM has the full support of all the relevant scientists and their governments. Since the governments in question include the administrations of George W. Bush, King Abdullah, John Howard and Hu Jintao, this is not a straightforward process; in fact there is something heroic about the firm stand the SPM manages to take. The price for this is that the SPM makes no policy recommendations of any kind, a fact which has drawn some negative comment; but the consensus on the basic facts is so remarkable that we can live without the unenforceable policy advice.

The first crucial component of the scientific consensus concerns a figure called the 'climate sensitivity'. This is the amount by which the climate will grow warmer if the amount of CO_2 in the atmosphere doubles. It is not a straightforward figure to calculate because many of the values change as the temperature changes; water vapour, for instance, is an important greenhouse gas, and as the oceans warm, water vapour in the atmosphere increases both in amount and in its greenhouse properties. Arrhenius thought that it would take three thousand years for our activities to double the level of CO_2, which in 1750, before the Industrial Revolution, was about 280 parts per million (ppm). By now the level is 379ppm and rising sharply. As the Chinese and Indian economies take off and global levels of CO_2 begin to rise even more quickly, it seems a racing certainty that we will achieve that level of doubled emissions some time this century; at which point the 'climate sensitivity' will become the most important number in the world. So the fact that according to the IPCC *'...an assessed likely range'* for climate sensitivity can now be given *'...for the first time'* is of more than academic interest. That figure is likely – between 66 and 90 per cent probable – to be between 2 and 4.5° centigrade. The best estimate is for climate sensitivity to be 3° centigrade. *'Values substantially higher than 4.5° centigrade cannot be excluded.'*

The consequences of this are listed pretty dryly in the report: cold days and nights will be warmer and fewer, hot ones hotter and more frequent – this is *'virtually certain'*, i.e. more than 99 per cent probable. Increased frequency of heatwaves and *'heavy precipitation events'* is *'very likely'* – 90 to 95 per cent. That means that a greater proportion of rain will come in the form of downpours. There will be more and bigger droughts, more and bigger tropical storms, and more and bigger floods – all *'likely'*, 66 to 90 per cent. The sea level will rise between 18 and 59 centimetres, mainly as a result of the ocean expanding as it warms. Increased melting in the Greenland and Antarctic is not included in these figures because there is not enough of a consensus to include its effects in the modelling. That isn't reassuring. The Greenland ice sheet holds enough water to raise global sea levels by seven metres – which would mean the end of, for instance, London, Miami, the Netherlands and Bangladesh.

What does the picture painted by the SPM mean? The short answer is that no one knows. Although we know more about many aspects of the climate than we once did, the fact is that we are entering a period of climatic change outside the experience of recorded human history, without a confident sense of what those changes will entail. If the events listed above are the whole of the story it doesn't seem too bad: hotter days and nights, storms and droughts, sound like things we should be able to endure. The trouble is that the global climate is a system of such complexity that we can't model in sufficient detail what the effects are. The last time

CO_2 levels were as high as they are today, in the last interglacial 125,000 years ago, sea levels were between four and six metres higher than they are today – a figure which we can take as a proxy for changes which in most respects are beyond imagining. What would happen if the harvest failed all across Europe or the US or Africa? What would happen if it failed again the next year, and the year after that? What would happen if the rain-and-meltwater pattern in the Yangtze valley, the core of Chinese agriculture, changed? What would happen if the glacial run-off from the Himalayas, which supplies most of India with its water, were to change? What would happen if the behaviour of El Niño were to become so unpredictable that agriculture in the Southern Hemisphere became unsustainable at current population levels? What would happen if those glaciers were to melt away? What would happen if the Gulf Stream (the Atlantic's 'meridional overturning circulation', as it is scientifically known) were to shut down suddenly – *The Day after Tomorrow* disaster scenario? The prediction is that Western Europe would become 8° centigrade cooler, about the temperature of Canada. But Canada produces enough food to feed 30 million people and enough grain to feed 60 million. Western Europe has a population of about 450 million. So what would they eat?[2] Hurricane Katrina gave us a glimpse of how quickly a meteorological event can destroy a city in the richest country in the world. We may be moving towards a future in which events like that come to seem commonplace. Anything in the paper today, darling? Not much – oh, all the Dutch drowned.

The SPM is silent about these horror scenarios, since its brief is to stick to certainties. It is also silent about the largest source of uncertainty, which as it happens doubles as the largest source of fear: feedback. Feedback is the process by which change increases further change: one example is that already mentioned, water vapour. A warmer world has more water vapour, and the warmer the vapour the more effective it is at blocking the escape of infra-red radiation, so the warmer the earth and oceans become, the more water vapour there is, and so on. Another feedback effect is caused by the melting of ice. Ice has a very high albedo – meaning, it reflects a lot of the sun's heat straight back out into space. This in turn has a cooling effect, which causes the manufacture of more ice, and so on. When ice melts, it turns into water, which has a much lower albedo, and therefore absorbs more heat, and therefore causes more ice to melt, which lowers the earth's net albedo even more, which causes it to be warmer, and so on. The wide variations in the earth's climate over geological timescales is thought to owe a great deal to feedback effects.

Some of these feedback effects are well known; when that is the case they have been included in the SPM. Others are a source of speculation and study but haven't yet reached a point of consensus where they can be included. One of the primary uncertainties is to do with clouds: no one knows how the changing climate will change the earth's cloud cover, and whether that will have a net cooling effect

2   About the only glimmer of good news in the SPM is that this event is rated 'very unlikely' to happen this century – less than 10 per cent probable.

(broadly speaking, low cloud blocks radiation from reaching the earth) or a warming one (equally broadly, high cloud prevents radiation from escaping into space). For that reason, the SPM is, for all its scariness, a conservative document: a lot of things that might be very bad news have been left out. Rising temperatures, even in the low and middle range of predictions in the SPM, can affect crop yields, kill off the Amazonian forest (according to a paper quoted by Monbiot, which says that *'...the Amazonian forest is near its critical resiliency threshold'* and risks becoming *'...essentially void of vegetation'*) and cause a shift in the fundamental ecological balance of the earth so that by about 2040 *'...living systems on the land will start to release more carbon dioxide than they absorb'*. As the permafrost in the Northern Hemisphere melts, it releases methane, a far more potent greenhouse gas than carbon; there are also enormous deposits of methane in the ocean, known as clathrates. The release of this methane on a large scale would hugely compound the effects of man-made greenhouse gases, and would lead to yet further greenhouse effects. Nobody knows when these feedback effects would kick in, but a number of climate scientists have said that they consider 2° centigrade to be the critical threshold. Because there is a time lag between the release of greenhouse gases into the atmosphere and the consequent rise in temperature, it is possible that we are already too late to avoid going through that 2° centigrade limit. That may in turn have committed us to climate change of a type that would destroy civilisation as we have known it. Humanity would be reduced to a small number of 'breeding pairs'. James Lovelock ends his book with a glimpse of what that might look like:

> *'Meanwhile in the hot arid world survivors gather for the journey to the new Arctic centres of civilisation; I see them in the desert as the dawn breaks and the sun throws its piercing gaze across the horizon at the camp. The cool fresh night air lingers for a while and then, like smoke, dissipates as the heat takes charge. Their camel wakes, blinks and slowly rises on her haunches. The few remaining members of the tribe mount. She belches, and sets off on the long unbearably hot journey to the next oasis.'*

There are three ways in which disaster might be avoided. First, the scientific consensus might be wrong. We might come out on the lucky end of that 90-95 per cent probability. Second, there might be some as-yet-undiscovered feedback effect that acts to counteract the warming caused by greenhouse gases. Recent years have benefited from the cooling effect of sulphur-based aerosols at high levels of the atmosphere (though that's about to come to an end, owing to legislation designed to cut pollution). We have also had the cooling effect caused by the explosion of Mount Pinitaubo in the Philippines in 1991. Perhaps a series of similar local cooling mechanisms will buy us some time, or perhaps there will be some huge new feedback that stops all the bad things from happening.

The third way disaster might be avoided is through action. Never before have we, on a planetary scale, so needed to combine pessimism of the intellect with optimism of the will. The pessimism is relatively easy to come by; the optimism less so. That is largely because our civilisation is based on the use of fossil fuels; they have been indispensable to the growth of material prosperity, a point made with great force by Richard Heinberg in his book *The Party's Over*. Fossil fuels underpin all of modern economic activity, science and technology, and at the moment there is no developed alternative for the future. Hundreds of millions of people in the developing world are starting to demand the very same First World lifestyle which has been bought by the use of those fuels.

The centrality of fossil fuels to our culture is the reason there has been no action on climate change since the UN agreed the Framework Convention on Climate Change in 1992. Fifteen years of inactivity ensued. (Even W. says so. *'Now is the time to act'*, he said in a speech in 2005. *'Now is the time to put a strategy [sic] – we should have done this ten to fifteen years ago.'*) The single item one can point to as an achievement is the Kyoto Treaty of 1997, which is worthless: even if fully implemented, Kyoto would have the effect of postponing the warming which would have been achieved without the treaty for all of six years, from 2094 to 2100. The reason for the inactivity is simple: we don't want to change. The prosperity brought by unchecked use of fossil fuels, and the concomitant economic growth of the past decade and a bit, are just too comfy-making. No politician is better informed than Al Gore about climate change, or more publicly identified with the necessity for action to combat it; but during his time as vice president, the US exceeded its stated targets for emissions by 15 per cent. Politicians are willing to talk about climate change but have as yet shown no willingness to act in any meaningful ways. In Britain, for example, climate change was for a period Tony Blair's preferred focus in his grotesque search for a 'legacy'. At the same time his government was committed to the largest expansion in airport traffic in British history, from 216 million passengers passing through in 2005 to a projected figure of 470 million in 2030. The government is building the runways to make sure that happens; and all this on top of the 500 per cent growth in UK air travel over the last 30 years. The government knows perfectly well that flying is the most potent way of emitting greenhouse gas – it magnifies the effect of those gases by 270 per cent. So why isn't it doing anything to stop the growth in flying? Won't this frenzied expansion trash any prospects of meeting targets to curb our emissions? Not a bit of it – because airline emissions aren't counted in the national figures. So from the government's point of view they don't matter.

It would be a mistake to see the orgy of runway-building as a contradiction of government policy in respect of global warming. In practical terms, there is no policy. Monbiot makes a horribly strong point in *Heat* (and graciously attributes it to his researcher Matthew Prescott):

> *'...government policy is not contained within the reports and reviews it commissions; government policy is the reports and reviews. By commissioning*

endless inquiries into the problem and the means by which it might be tackled, the government creates the impression that something is being done, while simultaneously preventing anything from happening until the next review (required to respond to the findings of the last review) has been published.'

The government is dedicating £3.6 billion to widening the M1, seven times what it is spending 'on policies that tackle climate change' – and if that last piece of ministerial wording fails to set off your bullshit detector, it's time to get the battery checked.

What is to be done? The first thing to do is to admit that Dick Cheney is right. *'Conservation may be a personal virtue,'* he said in 2001, *'...but it is not a sufficient basis for a sound, comprehensive energy policy'*. Rephrase that sentence to state that conservation is indeed a personal virtue, and both halves of it are, it seems to me, true. But there is also a problem with the notion of conservation as a personal virtue. The risk is that awareness of global warming and of the need to act to counter it can be reduced to a form of personal good conduct; to membership of the tribe of the virtuous. It is a good thing to choose to pollute less, to ride a bicycle and take the train and turn down the thermostat, and to fit low-energy lightbulbs, but there is a serious risk that these activities will come to seem an end in themselves, a meaningful contribution to the fight against climate change. They aren't. The changes that are needed are global and structural, and anything which distracts attention from that is potentially damaging. There is a parallel of sorts between militant conservationism and driving an SUV. The SUV driver is consciously choosing to worsen the environment, and to harm the planet, and is trying at the same time to send a signal – a signal to herself – that even if climate change comes she will be able to protect herself from it. Look, the huge car says: I can protect myself and my family, whatever happens. That is a falsehood, and it is a falsehood related to the idea that our individual choices are of any consequence. I've just switched my electricity supply to a green company. I did it to give myself the feeling that I'm doing what little I can. But this, too, is a kind of category mistake – the SUV driver isn't protecting anyone, and neither am I.

The second part of Cheney's observation is also true. A sound and comprehensive energy policy is what the world needs, with an important rider: that it is based on clean, non-carbon-producing sources of energy. It is over this question that the Bush administration has failed most spectacularly. The Bush congeries of oilmen and corporate shills had the worst possible motives for arguing that global warming could be addressed only through a technological solution; and the damage done over the Kyoto Treaty was not so much caused by the rejection per se (it had already been the subject of a 95-0 rejection by the Senate) as the subsequent refusal to engage with the subject, combined with an apparent attempt to suppress the scientific realities. But this doesn't mean that the administration was necessarily wrong about the crucial importance of as yet non-existent technologies to address global warming. There are two areas in particular where new technologies are needed. One is in the field of carbon capture and storage. These are techniques by which fossil fuels – coal or gas – are used to provide energy as they are now, but the CO_2 they produce is extracted from the emissions. Statoil, a Norwegian oil company (partly owned by the state), already sequesters emissions from gas it extracts from a North Sea oilfield; in Uzbekistan, a technique called gasification is used to turn CO_2 emissions into liquid and store them. There are problems with the question of storage – obviously, leaks are a seriously Bad Thing – but this is a technology with proven potential. It has a particular importance in the context of China, which is planning to fuel its future economic growth by building thousands of coal-fired power stations; coal being the dirtiest energy source of them all. So if we – the Western 'we' – were to develop an effective and cheap way of capturing and storing carbon emissions from these Chinese power stations we would in one move be making an important step towards controlling the planet's total future emissions. It stands to reason that the people who believe most sincerely in a technological solution to the problem of emissions would be backing a huge programme of investment in CCS. Except they haven't been. The rhetoric about technology has not been backed by the necessarily expensive and urgent action. The Internatonal Energy Agency (IEA) says that '*...large scale carbon capture and storage is probably ten years off, with real potential as an emission mitigation tool from 2030 in developed countries*'. That may well be too late.

One thing the Bushies are willing to talk about is hydrogen. This is, as its boosters like to point out, '*...the most common element in the universe*', which is true but irrelevant, since the hydrogen we can use is right here on earth, and at the moment takes more energy to extract than it supplies. It has potential, though, because it is practical and easy to use, and super-clean – its only emission takes the form of water. The technology is feasible and in its likeliest form would involve hydrogen boilers, hydrogen fuel cells (which are roughly analogous to batteries) and a new infrastructure of pipes to move the gas. In Monbiot's words, this would amount to '*...a massive and extremely ambitious government programme*', It is not what the Bushies have in mind, however, and their talk about hydrogen – sometimes, the 'hydrogen economy' – has behind it the moonshine of the big car manufacturers. By promising a super cool new kind of car in the middle distance, Detroit has managed to avoid the issue in the present, and continues its enthusiasm for the grotesquely bloated SUVs which both symbolise and enact an indifference to the rest of the planet. It is genuinely bizarre that just as people have become increasingly aware of the costs and consequences of fossil-fuel consumption, the mileage per gallon of American cars has actually gone down: 20.8 mpg, down from 22.1 in 1988, and this despite the fact that the cost of petrol has gone shooting up. A century ago, the Model T Ford did 25 mpg. Monbiot is quite right when he says that '*...it is beginning to look like the last days of the Roman Empire*'. Given these institutional pressures and the level of denial involved, it would seem unwise to wait for hydrogen to become a practicable energy source: we might have to wait a couple of decades, and we don't have a couple of decades. Bush is also touting ethanol, manufactured from corn, as a possible source of energy. This is popular with farmers and the corn-growing states, as you'd expect, even though it is potentially disastrous as a model for the rest of the planet, since the last thing we need is for even more forests to disappear.

The idea of getting enough power from renewable sources, wind and tide and the sun, is appealing; but the complexities involved in the management of power demand, as much as the sheer difficulty of generating the amounts needed, make me think it unlikely that the UK or anywhere else will ever come close to generating the power it needs from renewable energy. There is a contradiction, or a touch of wishful thinking, in much of the green polemics on this issue. They argue for the extreme urgency of the transition to low-emission energy, and at the same time for the importance of adopting experimental clean technologies. Those two things don't gel. Much of the green critique of existing arrangements is salutary – to take one tiny fact, it is astonishing that 66 per cent of the power generated by the UK national grid is wasted because of inefficiencies of transmission. But the green solutions – micro-generation, renewable energy and so on – even if not utopian, aren't right here, right now, where we need them. If we need the non-carbon-producing power urgently, and we do, then nuclear is for the moment our only choice – for the moment, meaning the next couple of decades. Monbiot goes into some detail about the potential for high-voltage DC transmission of power, which would enable energy to be moved over greater distances with little loss. This in turn would enable us to draw our energy from greater distances than before, and use offshore wind turbines and solar energy.

> *'For years, rogue environmentalists have been pointing out that solar electricity generated in the Sahara could power all of Europe, the Gobi could power China, and the Chihuahuan, Sonoran, Atacam and Great Victoria deserts could electrify their entire continents. These people have been dismissed as nutters. The development of cheap DC cables suggests that they might one day be proved right.'*

These are encouraging prospects; in the middle distance there is the potential for clean energy. In the meantime, unfortunately, it is likely that here in the UK we are going to need nuclear power. James Lovelock is a powerful advocate of this line of thinking, which is to me persuasive: the argument, put simply, is that nuclear power is a mature technology whose risks are understood, which would produce all the energy we need, and which is considered in the round the least worst solution to our urgent need for a carbon-free fuel source. It is not a prospect that brings much joy, and it is going to be of more than academic interest to see how the government gets round or forces its way past the inevitable local objections. We can all expect to hear a very great deal about how France gets 78 per cent of its electricity from nuclear power.

It all comes down to the question of political will. The remarkable thing is that most of the things we need to do to prevent climate change are clear in their outline, even though one can argue over details. We need to insulate our houses, on a massive scale; find an effective form of taxing the output of carbon (rather than just giving tradeable credits to the largest polluters, which is what the EU did – a policy that amounted to a 30 billion euro grant to the continent's biggest polluters); spend a fortune on both building and researching renewable energy and DC power; spend another fortune on nuclear power; double or treble our spending on public transport; do everything possible to curb the growth of air travel; and investigate what we need to do to defend ourselves if the sea rises, or if food imports collapse. If we do that we may find that we develop the technologies that China and India will need. If we can show that it is possible to cut carbon output dramatically without trashing our economy – well, that might be the single most important thing we could do, far outweighing the actual impact of our emission reductions.

We know all this, but whether any of it will actually happen is a different question. It is easy for politicians to stick wind turbines on their houses and ride bicycles, but effective action on climate change is about to require doing things that are not popular. In his eponymous report, Nicholas Stern has argued that it would cost about one per cent of global GDP now to prevent a loss of five per cent of global GDP in the future. The calculation is tweaked to make the cost now sound manageably small – but it is not yet clear whether Western electorates are willing to pay it. One per cent of global GDP is 600 billion dollars, most of which would be paid by the developed world. The idea is that by paying it now we would be keeping the world's economy on track so that by 2050 the developed world would be 200 per cent richer and the developing world 400 per cent, while our emissions decline by 60 to 90 per cent and theirs increase by 25 to 50. (One problem is that 17 per cent of that growth in developing world emissions has already been used up.) The promised economic growth is jam tomorrow; we would be paying for it today, in the form of increased taxes and lost jobs. These things are all real to voters in ways that climate change perhaps is not. Are people going to give things up in the present in order to prevent things that computer models tell them are going to happen in 25 years' time? If they – we – aren't, then we're heading for breeding pairs, and camels in the Arctic

This is an edited version of an article which first appeared in the *London Review of Books* – www.lrb.co.uk

Books referred to in this chapter

Frum, D. (2003) *The Right Man*, Random House

Heinberg, R. (2nd edition 2005) *The Party's Over: Oil, War and the Fate of Industrial Societies*, New Society Publishers

Intergovernmental Panel on Climate Change (2007) *Climate Change 2007: The Physical Science Basis Summary for Policymakers*, contribution of Working Group I to the Fourth Assessment Report of the IPCC: www.ipcc.ch/spm2feb07.pdf

Lovelock, J. (2006) *The Revenge of Gaia*, Allen Lane

Monbiot, G. (2006) *Heat: How to Stop the Planet Burning*, Penguin

Stanley Robinson, K. (2004) *Forty signs of rain*, Specta

Stern, S. (2006) *The Economics of Climate Change: The Stern Review*, HM Treasury

Other work consulted in the writing of this piece

Brown, P. (2006) *The Last Chance for Change*, Black

Dow, K. and Downing, T. (2006) *The Atlas of Climate Change: Mapping the World's Greatest Challenge*, Earthscan

Evans, K. (2006) *Funny Weather: Everything You Didn't Want to Know about Climate Change but Probably Should Find Out*, Myriad

Flannery, T. (2006) *The Weather Makers: The History and Future Impact of Climate Change*, Allen Lane

Guggenheim, D. (DVD, 2006) *An Inconvenient Truth: A Global Warning*, directed by Davis Guggenheim for Paramount

Kolbert, E. (2006) *Field Notes from a Catastrophe*, Bloomsbury

McGuire, B. (2006) *Catastrophes: A Very Short Introduction*, Oxford

Park, C. (2006) *Dictionary of Environment and Conservation*, Oxford

Sale, K. (new edition 2006) *After Eden: The Evolution of Human Domination*, Duke University Press

Steffen, A. (2006) *Worldchanging: A User's Guide for the 21st Century*, Abrams

Sweet, W. (2006) *Kicking the Carbon Habit: Global Warming and the Case for Renewable and Nuclear Energy*, Columbia

Worldwatch Institute (2006) *State of the World 2006: The Challenge of World Sustainability*, Earthscan

CHAPTER 3:
People and place – towards a culture of sustainability

Bruce McVean

Introduction

Despite the growing consensus amongst politicians, the business community and the public that we must do something to reduce our impact on the environment we're still a long way from achieving Robert Gray's classic definition of sustainable development as *'…treating the Earth as if we intended to stay'* (Tickell, 2007).

As the United Nations Environment Programme recently concluded, *'…the most recent data provide more evidence of growing pressures that damage the ecological systems supporting all life on the planet'* (UNEP, 2007). The overall trend is towards continued overuse of the Earth's ecosystem resources. Humanity's resource consumption and waste production exceeded the Earth biocapacity by about 25 per cent in 2003 (ibid.). The 2005 Millennium Ecosystem Assessment estimated that 15 of the 24 major ecosystem services that support humanity – through provision of fresh water, replenishment of fertile soil, or regulation of the climate for example – are being pushed to their sustainable limits or are already operating in a degraded state (ibid.).

While the recent interest in and increased awareness of sustainability is being driven by rightful concerns about the impacts of climate change, the concept of sustainable development is not a new one. It's over 30 years since the term was first popularised in the book, *Limits to Growth* (Meadows *et al.*, 1972) and over 20 years since what is still the standard definition of sustainable development *'development that meets the needs of the present without compromising the ability of future generations to meet their own needs'* – was first coined by the Brundtland Commission in its influential report, *Our Common Future* (World Commission on Environment and Development, 1987).

Almost a decade and a half ago the author and commentator James Howard Kunstler (1993) published his damning history of suburban development in the United States. Noting modern suburbia's impact on both the environment and society he argued that, *'…the joyride is over'*. What remains is the question of how we can make the transition to a saner way of living. To do so will certainly require a transformation of the physical setting for our civilisation, a remaking of the places where we live and work (ibid.). Of course, the UK has yet to achieve American levels of suburban sprawl and car dependence, but it is a sobering thought that Kunstler's lament would still apply to the design of the majority of the suburban and urban housing that is currently under construction across the country.

This begs the question of why, despite the fact that delivering sustainable development has been the primary function of the planning system since 2005 (ODPM, 2005), are there still no significant completed developments that could be described as genuinely sustainable?[1]

And why, as the Commission for Architecture and the Built Environment (CABE) concluded recently, are there so few even on the drawing board?

> 'We are now seeing many more projects coming through design review that aspire to reduce their environmental impacts and make a positive contribution to social and economic welfare. But these proposals often display only partial consideration of the issues' (CABE, 2007b).

At the time of writing there is only a handful of sizeable housing schemes and masterplans on the drawing board, in planning, or in existence which could be described as places designed for sustainability (ibid.).

The first part of the answer, as the quote from CABE suggests, is that there remains a tendency to consider sustainability as a separate issue rather than an integral part of everything we do. This is compounded by the fact that despite the rhetoric of 'triple bottom lines' and 'three legged stools', sustainability is still largely seen as an environmental issue.

This tendency to view sustainability in isolation leads to the second part of the answer; it allows us to believe that we can deliver a sustainable future by tinkering with business-as-usual, through bolt-ons and half measures. Instead, we must accept that we long ago moved beyond a point where minor alterations rather than fundamental changes are all that's required. As Thomas Homer-Dixon argues:

> 'The accelerating human transformation of the Earth's environment is not sustainable. Therefore the business-as-usual way of dealing with the Earth's System is not an option. It has to be replaced – as soon as possible – by deliberate strategies of management that sustain the Earth's environment while meeting social and economic development objectives' (Homer-Dixon, 2007).

Drivers for change

> 'Sustainability may be best defined as the capacity for continuance into the long-term future. Anything that can go on being done on an indefinite basis is sustainable. Anything that cannot go on being done indefinitely is unsustainable.

---

1   Planning issues are discussed in more detail in Chapter 5.

In that respect, sustainabaility is the end goal, or desired destination, for the human species' (Porritt, 2005).

There are a number of drivers putting sustainable development at the heart of all decisions relating to the design, construction and management of housing.

The Intergovernmental Panel on Climate Change (IPCC) fourth assessment report published in 2007, has underlined the serious and worsening impacts that unrestrained carbon emissions are having on the global climate. Impacts that were thought to be extreme and unlikely five years ago are now considered, in the light of the research undertaken over the last five years, to be extremely likely (in scientific terms, 90 per cent likely) to occur (IPCC, 2007). The 2006 Stern Review, the first authoritative analysis on the economic cost of climate change, made it very clear that the cost of not acting to restrict climate damage will be much higher than taking action to reduce carbon emissions (Stern, 2006).

The political and legislative response to this confluence of opinion, in favour of action to mitigate climate change impacts, is beginning to emerge. The current aspiration of government, to reduce carbon emissions by 60 per cent over 1990 levels by 2050, is almost certain to become a legally binding target in the Climate Change Bill and the target may even increase to a reduction in carbon emissions of 80 per cent by 2050.

We are already beginning to see stronger action being taken by the UK government. Changes to the planning system introduced since 2004 (including planning policy statements on climate change and renewable energy), the recent introduction of the Code for Sustainable Homes, higher building regulations standards for energy and soon water consumption and the proposal for all new homes to be zero-carbon by 2016, all show a readiness and commitment by government to take action to improve the sustainability performance of the built environment.

But we are unlikely to make the necessary changes quickly enough if all we do is respond to the latest policy. Housing associations have a proud history of rising to the social and environmental challenges of the time; of driving rather than merely responding to policy agendas. The social aims of housing associations and their ability to look beyond profits mean that they are ideally placed to push the sustainability agenda, and indeed projects such as BedZed have helped lay the ground for the current policy agenda (see Chapter 7 for a detailed discussion).

The creation and maintenance of sustainable places, places that enable residents to lead more sustainable lives, must then be the primary focus of the social housing sector. The physical and social environments we create are the product of thousands of conscious and unconscious choices, forces and processes. The design and management of these environments help make us well or ill, fit or obese, communal or individualistic, even happy or unhappy. Our urban environments – the human habitat – no less natural in most ways than an ant nest or sand dune for the flora and fauna it supports – are products of societies, their values and culture.

'We shape our buildings and afterwards our buildings shape us', so said Winston Churchill. And how much more true is it of whole places – neighbourhoods, towns and cities – than just buildings.

When it's done right the results are sublime. Take Kevin O'Sullivan's description of the Park Street Estate on London's South Bank as an example:

> *'With a physical layout that encourages connectedness, and a community culture that emphasises respect for difference, it provides what its residents need: homes from which they can pursue their purpose. In place of community...the estate offers both security and openness. There is a familiar and comfortable physical environment: neighbours, gardens, a place to park your car...And because the Park Street Estate is where it is, all the residents have within a short walk or a short ride on public transport all the offerings, facilities and community links of the metropolis'* (O'Sullivan, 2006).

Compare this with Holly Street Estate in Hackney, completed a little over a decade before Park Street:

> *'The design was completely alien to Hackney's principal residential urban form of grids of terraced streets alongside parks and squares...As its attractiveness declined and it became hard to let the problems of the tenants became commensurately worse. Just before redevelopment 31 per cent were unemployed, 21 per cent were lone parents and 63 per cent were on housing benefit. Some 80 per cent had applied to leave the estate. Crime and drug abuse were rife, with dealing commonplace in the brutalist concrete lobbies of the tower blocks'* (Simmons, 2006).

It is easy to dismiss the problems of estates such as Holly Street as just the result of the 60s and 70s hubris on the part of politicians, planners and architects. But are things really that different today?

> *'Across England, we have found that only 18 per cent – fewer than one in five – of developments we audited could be classed as 'good' or 'very good'. Perhaps more importantly, the quality of a substantial minority of developments – 29 per cent – is so low that they simply should not have been given planning consent... We also found that the poorest market housing schemes tended to be located in less affluent neighbourhoods'* (CABE, 2007a).

There is no easy explanation as to why this should be so. Arguably there is plenty of evidence as to how to create sustainable places and buildings. And as the Royal Commission on Environmental Pollution (2007) has concluded we are not dependent upon extraordinary new technological innovation, '*...when it comes to improving the urban environment, many techniques are readily available and have been shown to work in demonstration projects, but have not become standard practice*'. And yet good practice in the UK remains elusive.

The environmental imperative

'*The twenty-first century will, in fact, be the Age of Nature. We'll learn, probably the hard way, that nature matters: we're not separate from it, we're dependent on it, and when there's trouble in nature, there's trouble in society*' (Homer-Dixon, 2006).

Sustainable development is not simply about managing the environment. Social and economic objectives matter as much as environmental, with a core aim being to optimise wellbeing. But, and it is a *big* but, we must recognise in pursing social and economic goals that we operate within environmental limits.

As good a place to start as any, given the urgent need to address man-made climate change, is accepting the need to cut carbon emissions by at least 80 per cent by 2050. Climate change will be this century's defining environmental challenge. Every action should be assessed with regard to its potential to contribute to climate change. That is not to say that cutting carbon emissions needs to be favoured over social and economic issues.

Rather, our efforts to deal with climate change should be seen as an opportunity to also address a wide range of other pressing environmental, social and economic concerns. For example, efforts to reduce carbon emissions from private motor car use will improve urban air quality, help combat rising levels of obesity and reduce the significant social and economic costs of accidents and congestion. Improving the energy efficiency of existing stock, a huge challenge given that the UK's building stock is currently replaced at around one per cent a year, will help tackle fuel poverty and reduce the respiratory and other illnesses associated with living in cold and damp accommodation (see Chapter 10).

So, the primary guiding principle for the design and management of new and existing places must be to accept the environmental imperative, to learn to live sustainably within the limits of the natural systems that provide the foundation for life. If we don't, as Porritt argues:

'*...then we will go the same way as every other life form that failed adapt to those changing systems and limits...If we can't secure our own biophysical survival, then it is game over...With great respect to those who assert the so called 'primacy' of key social and economic goals it must be said loud and clear that these are secondary goals: all else is conditional upon learning to live sustainably within the Earth's systems and limits. Not only is the pursuit of biophysical sustainability non-negotiable; it's preconditional*' (Porritt, 2005).

Towards a culture of sustainability

'Ultimately, it is certain patterns of human behaviour that lead to environmental degradation, and other patterns that result in sustainable development' (Silver and Defries, 1990).

Genuine sustainable development cannot be achieved simply by relying on technological advances or physical interventions. For example, the former Mayor of London has stated that 18 per cent of domestic carbon reductions will be met through *'behaviour change'* (GLA, 2007), and research by Sustrans (2007) has found that factors such as lack of information or negative perceptions result in between a quarter and a third of journeys being made by car rather than on foot, by bike or public transport.

Housing providers must therefore consider how the places they are responsible for can be designed and managed to encourage people to live sustainably. To walk, cycle and use public transport, to value public spaces and use local shops, to recycle, to save water. Some of this can be very obvious, with signage or live monitors showing energy use, but on the whole it is implicit in how a place is laid out and constructed. This is the difference between a housing development that aims to be truly sustainable, and one that seeks merely to mitigate some of its most serious impacts.

The key question is how can we make the sustainable choice the obvious choice? A sustainable place is a composite of numerous physical and social elements, from streets, parks and homes to governance and ongoing maintenance and management regimes. None of these elements can be considered in isolation, approaches to design and management must seek to integrate all the elements with a view to enabling a culture of sustainable living among residents.

The success of developments of Hammarby Sjostad, (www.hammarbysjostad.se/) the new suburb in Stockholm that has become the byword for sustainable place making, show the benefits of taking an integrated approach. The aim was to halve the environmental impact when compared to a development built to Swedish building regulations (which are already much tougher than the UK's, particularly with regard to energy consumption). This was only achievable by taking an integrated approach to energy, waste, water, transport, building design and construction management. The result is not only high environmental performance of individual buildings, but also a new suburb with a high quality public realm, shops and other community facilities within walking distance, and excellent links to the city centre by public transport or bike. As David Birkbeck from Design for Homes commented, *'If there was an Olympics for sustainability, Sweden would top the medal table, and Hammarby would break all the records'* (Building Design, 2007).

When discussing comparative rates of walking and cycling, recycling or home energy generation between British and other European cities, a British audience will readily resign itself to the fact that the residents of Basel or Copenhagen, say walk, cycle and recycle more *because* they are Swiss and Danish! It's as if the propensity to walk, cycle and recycle is a genetic endowment rather than a construct of urban design, traffic planning, public policy and culture. If we want to influence human behaviour – especially everyday lifestyle decisions or choices – to be more sustainable we need to design and maintain our urban environments differently. And if and as we do this we will in turn shape behaviour until more sustainable living becomes progressively more enjoyable, more accessible, and more affordable.

Ultimately a virtuous cycle is set in motion and we end up with more humane, democratic and environmentally sustainable cities like Copenhagen in Denmark; Malmo in Sweden and Freiburg in Germany. The new environments we make then give rise to more humane, democratic and sustainable cultures and societies.

Creating those environments, enabling sustainable lifestyles is the key challenge for the housing sector. The following chapters will go a long way to helping you and your organisation rise to that challenge, but there are no easy answers or magic bullets. Tough questions will need to be asked, of everyone involved, but in the end it will be worth it.

References

Building Design (2007) 'Brown "inspired" by Swedish eco-town', *Building Design*, 1 June. Available from: http://www.bdonline.co.uk/stroy.asp?storycode=3088296

CABE (2007a) *Housing audit: Assessing the design quality of new housing in the East Midlands, West Midlands and the South West*, Commission for Architecture and the Built Environment. Available from: http://www.cabe.org.uk/default.aspx?contentitemid=1727

CABE (2007b) *Sustainable design, climate change and the built environment*, Commission for Architecture and the Built Environment. Available from: http://www.cabe.org.uk/default.aspx?contentitemid=2077

GLA (2007) *Action Today to Protect Tomorrow: The Mayor's Climate Change Action Plan*, London: Greater London Authority

Homer-Dixon, T. (2006) *The Upside of Down: Catastrophe, Creativity and the Renewal of Civilisation*, London: Souvenir Press

IPCC (2007) *Climate Change 2007: Summary for Policymakers,* Intergovernmental Panel on Climate Change. Available from: http://www.ipcc.ch/pdf/assessment-report/ar4/syr/ar4_syr_spm.pdf

Kunstler, J. H. (1993) *The Geography of Nowhere*, New York: Simon and Schuster

Meadows, D. H., Meadows, D. L., Randers, J. and Behrens III, W. W. (1972) *The limits to growth*, New York: Universe Books

ODPM (2005) *Planning Policy Statement 1: Delivering Sustainable Development*, London: Office of the Deputy Prime Minister. Available from: http://www.communities. gov.uk/publications/planningandbuilding/planningpolicystatement

O'Sullivan, K. (2006) 'A Niche in the Metropolis' in A. Buonfino and G. Mulgan, (2006) *Porcupines in Winter*, London: The Young Foundation

Porritt, J. (2005) *Capitalism as if the world matters*, London: Earthscan

Royal Commission on Environmental Pollution (2007) *The Urban Environment: Summary of the Royal Commission on Environmental Pollution's Report*. Available from: http://www.rcep.org.uk/urbanenvironment.htm

Silver, C. S. and Defries, R. S. (1990) *One Earth, One Future*, Washington: National Academy Press

Simmons, R. (2006) 'The cost of bad design' in CABE, *The cost of bad design*, London: Commission for Architecture and the Built Environment. Available from: http://www.cabe.org.uk/AssetLibrary/8125.pdf

Stern, N. (2006) *The Economics of Climate Change: The Stern Review*, London: HM Treasury. Available from: http://www.hm-treasury.gov.uk/independent_reviews/stern_review_economics_climate_change/sternreview_index.cfm

Sustrans (2007) *Travel Behaviour Research and Evaluation*. Available from: http://www.sustrans.org.uk/default.asp?sID=1173361630250

Tickell, C. (2007) *Sustainability and Conservation: Prospects for Johannesburg*. Available from: http://www.crispintickell.com/page20.html

UNEP (2007) *GEO Year Book 2007: An Overview of Our Changing Environment*, Nairobi: UNEP

Wheeler, S.M and Beatley, T. (Eds) *The Sustainable Urban Development Reader* London: Routledge

World Commission on Environment and Development (1987) *Our Common Future*, Oxford: Oxford University Press

CHAPTER 4:
Taking the lead – local authorities and the climate change agenda

Joanne Wade

Introduction

Local authorities can and should lead their local communities in responding to the challenge of climate change. Indeed, sustainable communities can only be shaped successfully if meeting the climate challenge is at the heart of local government's community leadership role. Action to reduce carbon emissions will only happen if people see the link between small, individual actions and the global problem of climate change, and this is where the involvement of local authorities is crucial.

In a recent report on how we can best engage people in tackling climate change, the Institute for Public Policy Research (2007) concludes that if people can see that climate change will have an effect in their neighbourhood, they are more likely to be able to understand the problem than if it is presented as some abstract global phenomenon. Also, people may be more motivated to tackle a large problem if they feel part of a communal response, as they are more likely to see that their actions could have an impact. Local authorities are in a key position to communicate the potential local impacts of climate change and bring the community together to develop a collective response.

All sectors of the community will need to be involved in implementing an effective response to climate change. New homes are to be zero-carbon by 2016, and should also be designed with a changing climate in mind; 'place shaping' needs to build communities that work in a low-carbon world and are resilient to climate change impacts; people must be encouraged and enabled to improve the quality of existing homes and helped to develop low-carbon lifestyles. Within their climate change leadership role, councils can have a positive impact in each of these areas, even though the overall task is potentially a huge one.

National policy seeks local leadership on climate change

With the introduction of the Climate Change Bill, the UK government is committing the country to taking action to address the challenge of climate change. If the bill is enacted as it stands, the country will be setting out on a path to reduce carbon dioxide emissions by 60 per cent by the year 2050.

There is no requirement in the Climate Change Bill for local authorities to reduce carbon emissions. However, other recent government policy on climate change and sustainable energy, set out in devolved administration and UK-level climate change programmes and the energy white paper, has emphasised the need for local authorities to address the issue. With the review of the local government performance framework in England, this contribution has become a central part of local authorities' local leadership role. It is likely that this shift in emphasis will also occur in the other parts of the UK in the near future.

Local government policy

In England, Scotland and Wales, local authorities have responsibility for the development of community plans.[1] In Northern Ireland there is currently no requirement for the preparation of community plans, although it is being considered as part of the ongoing Review of Local Government.

As explained in the planning chapter (Chapter 5), the aim of community plans is to secure the economic, social and environmental wellbeing of the local area, and government guidance to local authorities emphasises the importance of ensuring that they contribute to sustainable development.

The local government white paper (CLG, 2006) introduced the new performance management framework for local authorities in England, and also emphasised the importance of local authority leadership in tackling climate change (Annex F of the white paper). In this annex, the role of a local authority is seen as four-fold:

- offering strong and visible leadership;
- setting an example;
- responding to the priorities of the local community; and
- co-ordinating new delivery partnerships.

The way that a local authority's performance is measured is changing, with the introduction of Comprehensive Area Assessments (CAAs) that look at how an area is improving and whether there are risks to future improvements in wellbeing in that area. The national indicator set to be used in the CAA (HM Government, 2007) includes four indicators that relate closely to tackling climate change:

- carbon dioxide emissions reduction from local authority operations (NI 185);
- per capita carbon dioxide emissions reduction in the local authority area (NI 186);
- fuel poverty (NI 187); and
- planning to adapt to climate change (NI 188).

---

1   Known as Sustainable Community Strategies in England, Community Strategies in Wales and Community Plans in Scotland.

Local authorities will choose up to 35 of the 198 indicators to include in their Local Area Agreement (LAA), which forms the basis of their improvement plan for their area and will be at the core of the area risk assessment carried out by the Audit Commission.

Reductions in a local authority's own carbon dioxide emissions will be considered as part of the 'use of resources' element of the CAA, as well as through reporting against indicator 185. The present and likely future carbon performance of service delivery will be included in the area risk assessment if a local authority chooses to include indicators 186, 187 and/or 188 in its LAA.

The recently published report of the Local Government Association's Climate Change Commission (LGA, 2007) demands that all authorities see climate change mitigation and adaptation as central to their role, and the Department for Environment, Food and Rural Affairs (DEFRA) wants all local authorities to include NI 186 in their LAAs. Analysis commissioned by DEFRA (AEA Energy and Environment, 2007) has suggested that a reduction in emissions of between 11 and 13 per cent (between 2004 and 2010) could be aimed for in many local authority areas.

In Wales, the 2006 report of the Beecham review of local service delivery made a number of proposals relating to public service management and delivery. The Welsh Assembly Government (WAG) response to this included proposals for Local Service Boards and Local Service Agreements. These are the mechanisms for delivering community strategies in Wales.

In 2007's WAG consultation, climate change is defined as one of four underlying principles that should be considered in the development of community strategies.

A concordat between Scottish Government and local authorities in Scotland,[2] agreed in November 2007, links central funding for local authorities for the period 2008-2011 to the delivery of a set of national targets. Included in these is a reduction in carbon emissions over the period, with a longer-term target of an 80 per cent emissions reduction by 2050.

Planning, building regulations and housing

The planning system is seen as one of the main mechanisms through which the country can move towards a low-carbon future. Indeed, the recently published Planning Reform Bill (CLG, 2007) will, if enacted, place at the heart of Local Development Plan Documents the need to ensure that development and land-use contribute to climate change mitigation and adaptation. Local authorities in England, Scotland and Wales are responsible for planning policy at the local level, and the Review of Public Administration is considering transferring responsibility for planning

2 www.scotland.gov.uk/Resource/Doc/923/0054147.pdf

to local authorities in Northern Ireland. This topic is covered in detail elsewhere in Chapter 5, but it is worth here considering briefly the implications for local authorities of the move towards zero-carbon developments.

The government published *Building a Greener Future* (CLG, 2007b) in July 2007. This confirmed the government's intention that all new housing in England should be zero-carbon by 2016. WAG is following a similar path in Wales, and has announced an aim for all new developments (not just housing) to be zero-carbon by 2011. In Scotland a slightly different approach is being taken, with an independent expert panel on energy efficiency considering how building regulations in Scotland should be altered to achieve Scandinavian levels of energy efficiency in new housing. These policies are discussed further in the planning chapter, but it is worth noting that implementation – for example in England via successive tightening of the building regulations linked to the achievement of different levels of the *Code for Sustainable Homes* (CLG, 2006b) – has implications for the local authority planning and building control functions. Successful implementation will depend on local planning documents that are consistent with the requirements of the code, and also on robust regulation through building control.

Local authorities in England, Scotland and Wales have a range of other housing responsibilities, many of which link to efforts to tackle climate change. For example, the housing fitness or quality standards in all parts of the UK include elements relating to the thermal properties of dwellings and the efficiency of their heating systems.

All parts of the UK have strategies for tackling fuel poverty: the energy-efficiency improvements that occur as a result will not only increase access to affordable warmth but will also lead to reduced carbon dioxide emissions. As the introduction this year by DEFRA of a one-year Community Energy Efficiency Fund in England illustrates, local authorities are seen to have a crucial co-ordinating role in ensuring that the money available for tackling fuel poverty reaches those who can benefit from it.[3] However, there are obstacles at present as the data needed for effective co-ordination are not often made available to local authorities. This area of energy activity is discussed in more detail in Chapter 6 (on fuel poverty).

As well as requiring action from local government, national policy can also reinforce and support it. The European Energy Performance of Buildings Directive is leading to the introduction of Energy Performance Certificates for all buildings in the UK.[4]

---

3 The fund was introduced by DEFRA as the mechanism for allocating money committed by the Chancellor in the 2006 pre-Budget report, to improve the effectiveness of national fuel poverty and energy supplier energy efficiency programmes.

4 For more information on Energy Performance Certificates and their inclusion in Home Information Packs, see: www.homeinformationpacks.gov.uk/industry/69_Housing_associations_and_local_authorities.html

This is discussed in detail in Chapter 10 on existing housing, but it is worth noting here two potential implications for local authorities. First, the energy labels that it will introduce should over time increase awareness of energy efficiency in buildings and therefore should lend support to local authority efforts to engage the community in reducing carbon emissions at home and at work. Second, these certificates must be displayed in public buildings, and will thus also provide a mechanism through which local authorities can publicise efforts they are making to take the lead by improving energy efficiency in their own buildings.

The Local Government Association's Climate Change Commission

During 2007, the Local Government Association convened a Climate Change Commission, to consider how local government can respond more effectively to the challenges of carbon emissions reduction and adaptation to climate change. As mentioned previously, the Commission report (LGA, 2007) demands that local authorities prioritise action on climate change. The Commission wants all local authorities to be performing well on climate change by the end of 2009, with mitigation and adaptation included in all LAAs. One hundred per cent of authorities signed up to a climate change declaration, published climate change strategies and action plans, and local authority chief executives have been designated as carbon accounting officers. The Commission suggests that, if this does not occur, central government should act to introduce a statutory duty for local authorities to tackle climate change.

How will national policy develop in the future?

The LGA Climate Change Commission report (ibid.) recommends that the present statutory framework supporting local government action on climate change is strengthened in a number of ways including:

- adding to the Climate Change Bill a power to apply a new duty on all public bodies to tackle climate change;
- ensuring that national carbon emissions reduction targets are a material consideration for planning purposes; and
- allowing councils to fast-track carbon reduction policies within their local development frameworks.

A report for the Co-operative Bank and Friends of the Earth (Boardman, 2007), also published in November 2007, considers the additional policies needed to achieve an 80 per cent reduction in carbon dioxide emissions from housing. These are examined in more detail in Chapter 10 on existing housing.

Many of the recommendations in the report (such as tighter minimum efficiency standards for *all* housing, and financial support for investment in low-carbon technologies) would support local government actions to reduce emissions by providing necessary drivers at a national level. Others (in particular the suggested legally binding targets for local authorities for the energy efficiency of housing in their area) would help to force less forward-thinking local authorities to act. The extent to which national government will accept and implement any of them is unknown, but they give an indication of the sorts of things that may occur in the foreseeable future as efforts to reduce carbon dioxide emissions are stepped-up.

Adapting to climate change

People will need to be able to live comfortably in a changed climate. Local government's place shaping role therefore has to include improving resilience to the climate change that is unavoidable as well as mitigating against further, avoidable change. Here again, the planning system has a crucial role to play in ensuring that individual buildings are designed for a changed climate and in considering overall development design issues, such as the use of sustainable urban drainage systems to reduce flood risks.

The definition of national indicators (CLG, 2008) suggests that progressing beyond the first level of achievement against the adaptation indicator will involve a comprehensive local risk assessment of vulnerabilities to climate, identification of the most effective adaptation responses, linking these to council strategies and plans, developing an action plan based on these, and implementing this plan along with a monitoring and review process.

The UK Climate Impacts Programme has developed an evaluation process for local authorities that should help them to respond to this issue. *The Local Climate Impacts Profile* (Metcalf and Greenhalgh, 2007) offers a relatively simple way for local authorities to gain a better understanding of local exposure to both weather and climate. Based on the fact that extreme weather events offer the greatest potential threat to local communities, it focuses on these events and how the council would need to react to them. The development of the profile is a four-stage process. The first stage involves researching past extreme weather events, the impact these had on the community and the actions the council took in response. The second stage involves reacting to this information: making changes to procedures where necessary and developing an understanding of the thresholds beyond which past events have occurred (for example, how much intense rainfall is needed to result in flooding). This second stage also includes the setting up of monitoring to ensure that information about future events is routinely captured. Stage three involves employing (internal) expertise to validate the thresholds and then relating these to climate projections for the region: in other words translating the average climate projections into the likelihood of extreme weather events. This information can then be used as the basis for making decisions on adaptation strategies. A fourth stage of the process can be undertaken as necessary, expanding the initial historical data-gathering over a longer time period to generate a more sophisticated understanding of the impacts of weather events where these have been significant.

Improving local wellbeing

Historically, the scope for local authorities to act to tackle climate change would have been limited, since it was restricted to activities for which local authorities had specific powers. However, the Local Government Act 2000 (for England and Wales) and the Local Government (Scotland) Act 2003 removed many legal barriers to action through the introduction of the power of wellbeing. Introduction of a similar power for local authorities in Northern Ireland is being considered.

The power of 'wellbeing' is intended as a power of first resort, and enables a local authority to undertake any activity, unless expressly forbidden elsewhere in legislation, that is intended to improve the economic, social or environmental wellbeing of all or part of its area.

Action to reduce carbon dioxide emissions and therefore tackle climate change clearly contributes to environmental wellbeing. A brief look at the potential impacts of climate change makes clear that it can also contribute to social and economic wellbeing. The UK Climate Impacts Programme predicts a range of social and economic effects of unchecked climate change: water shortages may become more common; whilst on the other hand flood plains and coastal areas may be inundated more frequently; habitat loss may impact on biodiversity and on the tourism industry; excessive summer heat in cities will cause health problems, and storm damage and subsidence will have significant cost implications for the insurance industry.

The wellbeing power has not yet been extensively used. Some local authority officers remain unaware of its potential and many others are unsure how to best use it. However, a number of authorities have employed the power to support or enable their sustainable energy activities.[5] For example, Braintree District Council used it to support the provision of grants for renewable energy installations and using the power enabled Fenland District Council to participate in the development of an energy services company for households in their area. The most innovative use of the power to date is by Kirklees District Council: they have been able to join the UK Emissions Trading Scheme, based on the argument that this will contribute to local wellbeing.

Climate change in community strategies

Community strategies are intended to fulfil the duty of local authorities to set out how they will improve economic, social and environmental wellbeing in their areas. Including climate change aims within them provides an appropriate focus on the topic in the local area, helps to engage partners through the Local Strategic

---

5   More information on the use of the power, including the examples given here, can be found in the report: *Local authority legal powers to promote sustainable energy* and its accompanying case studies. Both are available at: www.impetusconsult.co.uk/publications.htm

Partnership (LSP), Community Planning Partnership or Local Service Board, offers a simple justification for the use of the power of wellbeing should that be appropriate, and links to including reduced carbon emissions as a target in the Local Area Agreement.

The Improvement and Development Agency for local government (IDeA) offers guidance on how climate change can be integrated into a community strategy.[6] This highlights the need for awareness of the issue of climate change within the partnership responsible for developing the strategy, and therefore the importance of ensuring that there is at least one sustainable development 'champion' on the partnership's board.

Climate change can be a separate theme for action, with its own task group, or can be integrated into existing environmental or sustainable development themes.

Whilst inclusion of climate change goals in the community strategy has benefits, as mentioned above, there may be limits on the extent of its influence. Partners included in the delivery of the strategy are likely to increase their level of activity, but whether organisations and individuals outside the partnership take action will depend on how the strategy is implemented.

Community engagement

The development or revision of a community strategy is an excellent focal point around which a local authority, working with its partners, can increase community understanding of the local implications of climate change and the opportunities to be involved in for acting to reduce emissions.

For example, Mid Bedfordshire District Council's LSP organised a one-day climate change conference,[7] to raise awareness and also to provide the LSP with access to the ideas and expertise of other members of the local community. The authority also took the opportunity to publicly sign the Nottingham Declaration, demonstrating their commitment to addressing climate change.

Existing local authority activities can contribute to tackling climate change

Many of the activities that local authorities already perform can contribute to reducing carbon dioxide emissions. Planning, housing and transport responsibilities are covered in other chapters of this book, and some key elements have been highlighted earlier in this chapter.

---

6   www.idea.gov.uk/idk/core/page.do?pageId=80857
7   www.community.midbeds.gov.uk/council/corporate/lsp/climate/default.asp

As yet, relatively few authorities have built on these foundations to take the lead in developing low-carbon communities. However, the majority have taken a first step, signing the Nottingham Declaration, the Welsh Declaration on Climate Change and Energy Efficiency or Scotland's Climate Change Declaration. These declarations are intended to demonstrate a local authority's commitment to leading local progress towards national mitigation and adaptation goals. They include commitments to tackling emissions from the local authority's own operations, encouraging action in the wider community, adapting to climate change and monitoring and reporting progress in these areas. Signing the relevant declaration is a good first step, but genuine progress can only be achieved if the commitments are implemented through firm action and community leadership.

Climate change strategies and action plans

Increasing numbers of authorities are following their initial commitment with the development of a strategic approach to the issue. Some have climate change action as a central community strategy aim. For example, the London Borough of Camden aims to be a 'low-carbon, low-waste' borough by 2012 (Camden, 2007) and has set a target of reducing carbon dioxide emissions in the borough by 60 per cent by 2050 (Camden, 2006). This aim can be supported by a climate change strategy and/or action plan: Aberdeen City Council was one of the earliest to develop such an approach, writing a climate change action plan in 2002.[8] The action plan focuses on encouraging and enabling community action, and includes provision of information to individuals, their communities and local businesses. There is a particular emphasis on working with local schools to make sure that climate change is included in the curriculum for 5 to 14 year olds. The plan is complemented by a carbon management implementation plan that addresses the council's own carbon emissions.

Leading by example

Aberdeen was one of the first local authorities to work in partnership with the Carbon Trust's Local Authority Carbon Management Programme to develop a corporate approach to carbon management.[9]

Local authorities wishing to inspire community action on carbon emissions need to be able to show that they are themselves taking the problem seriously and are prepared to act on their concerns. Developing and implementing a carbon management plan is one key way to do this. Using the communication opportunities offered by the implementation of Energy Performance Certificates for council buildings (see earlier) is another.

---

8   See: www.aberdeencity.gov.uk/ACCI/web/site/CommunityAdvice/SL(YourEnvironment)/cma_youenviron_climate.asp for more on climate change activities in the council and www.energysavingtrust.org.uk/housingbuildings/casestudies/index.cfm?mode=listing&audtype=2&casecat=76 for Energy Saving Trust case studies on this and other local authority strategies.
9   See: www.carbontrust.co.uk/carbon/la/ for more information about this programme.

Zero-carbon housing

An increasing number of local authorities are implementing planning policies that reduce carbon emissions from new developments, as discussed in detail in the planning Chapter 5.

Whilst many existing homes cannot sensibly be upgraded to become zero-carbon, there is much that can be done to reduce the level of emissions from these properties. There are lots of things that local authorities can do to encourage householders to reduce emissions, some of which are described below.[10]

Energy services

One effective way to help households invest in energy efficiency is to provide them with an affordable and simple source of finance for the investment. Offering an energy services[11] package is one way to do this. This may involve providing a loan for energy-efficiency measures or heating system improvements and then linking loan repayments to reductions in energy bills. The repayments can be collected alongside payments for fuel so that the householder receives one monthly bill for the overall service. The Greater London Energy Efficiency Network (GLEEN) has been running this type of scheme for a number of years through its energy service company HelpCo.[12]

Other options for energy services schemes include arranging affinity deals with energy suppliers. One example of this type of scheme is the Wasteless Society's Green Energy scheme,[13] where people can sign up to a renewable energy tariff from the scheme's preferred supplier who, in return, pay a commission to the Wasteless Society which it then uses to support energy-efficiency work.

Energy services approaches can support increased community involvement in the renewable supply and conservation of energy, and can offer financial benefits to communities beyond those associated with energy saving. For example, Gigha Renewable Energy Ltd[14] manages a set of three wind turbines on the island of Gigha, for the benefit of the island's whole community. The turbines are owned by the community and profit from sale of the electricity generated will be spent on investments to benefit the community, including energy saving measures in housing stock owned by the company's parent charity, the Island of Gigha Heritage Trust.

---

10 This issue is also addressed in Chapter 10 on existing housing.
11 Energy services are the things we use energy to achieve – for example a warm home or chilled food. Packages offered to consumers that include fuel and the equipment needed to transform it into the efficient delivery of an energy service (e.g. a heating system or home insulation) are referred to as energy services packages.
12 www.gleen.org.uk/what_does.html
13 www.wasteless.co.uk/html/greenelectricity.htm
14 gigha.org.uk/windmills/TheStoryoftheWindmills.php

There are a number of possible roles for local authorities in the development and operation of energy services schemes, and more information on this can be found in the Energy Saving Trust's Directory of Energy Services.[15]

Inspiring community action

The majority of people in the UK recognise climate change as a serious issue, and general attitudes are becoming more positive towards a need to take action. However, there remains a lot to be done. The Energy Saving Trust's 'Green Barometer' (EST, 2007) notes that one third of people admit to taking the car rather than walking for short journeys, over a third leave the TV on standby and a quarter regularly leave the lights on when leaving a room.

And it is not just our day to day behaviour that lags behind our attitudes; many of us are not yet investing in cost-effective energy-efficiency technologies for our homes. For example, the English House Condition Survey[16] reports that in 2004 there remained over 9 million homes in England with un-insulated cavity walls and 6.5 million with lofts that had less than 100mm of insulation.

Thus there is a need to translate attitudes into activity. How can local authorities inspire action? Perhaps by taking a lead or perhaps simply by supporting the work of motivated individuals within the community.

Taking the lead can mean working at a strategic level and helping entire communities to plot a path to lower carbon emissions. For example, parish councils in West Sussex are being helped by the 'LIVE!' project to develop dedicated sustainability groups within their communities that will work together to bring about increased action to reduce the environmental impact of the community.[17]

Alternatively, councils can work with partners to develop very specific initiatives that raise the profile of energy efficiency and entice householders to act. For example, Fenland District Council was the first authority in the country to link investment in energy efficiency to a council tax credit. Initially the scheme offered three different levels of credit linked to the extent of energy-efficiency improvement made to a home. More recently, a simplified scheme offering a £50 credit linked to installing specific measures has been introduced. This latter scheme involved partnership and funding from an energy supplier, and is an option that has been taken up by a number of other authorities too.[18]

---

15 Available at: www.energysavingtrust.org.uk/uploads/documents/housingbuildings/ESDirectory.pdf
16 www.communities.gov.uk/housing/housingresearch/housingsurveys/englishhousecondition/
17 See: www.impetusconsult.co.uk for more on this project.
18 For more on Fenland's activities see: www.fenland.gov.uk/ccm/navigation/environment/energy-efficiency/

A local council may also decide to invest directly in carbon emissions reduction in the community: the London Borough of Islington announced in February 2007 the establishment of a £3 million Climate Change Fund. This money is available to residents, community organisations and the council itself, and is intended to provide partial funding for renewable energy installations and innovative transport projects.

Increasingly, action is being led by highly motivated individuals and groups within local communities. The Transition Towns[19] initiative is a key example of this type of approach, in which the local community comes together to work towards a low-carbon future. At the time of writing there were 24 transition 'towns' registered in the UK. In some of these, local councillors were taking an active part in supporting and promoting the initiative.

Ensuring that people receive the advice they need

Information in a useful and accessible form is one of the key requirements preceding action. The more that this information comes in the form of tailored advice of direct relevance to the individual, the more likely it is to lead to action. As mentioned at the beginning of this chapter, local authorities are well positioned to communicate effectively with local communities about the problem of climate change and its potential solutions and, as part of this, they can signpost people to the advice that they need.

Advice provision is developing rapidly at the present time. For a number of years the Energy Saving Trust has funded a network of Energy Efficiency Advice Centres across the country and these have had some success in helping people to improve the energy efficiency of their homes. These are now being restructured and their remit expanded so that more proactive and personalised advice can be provided. Alongside this, a new paid-for service will be launched that will provide a hands-on, hassle-free way for people to reduce carbon emissions and other negative environmental impacts from their homes. This 'Green Homes Service' will be launched during 2008, and local authorities should see it as a potentially very useful partner in their efforts to encourage people to do more.[20]

The introduction of Energy Performance Certificates (EPCs) for homes offers a new hook from which carbon emissions reduction advice can be hung. In particular, councils could look to working with estate agents and linking the offer of advice to home purchase or new rentals as this is when the occupant would be most aware of the information in the EPC.

---

19 For more information on the Transition Towns initiative see: transitiontowns.org/
20 For up to date information on the Green Homes service, see:
www.energysavingtrust.org.uk/help_and_support/green_homes_service

Barriers to local leadership on climate change

The preceding sections of this chapter have given a few examples of the many things that local authorities can do and are doing to lead their communities in tackling climate change. However, there remain many authorities that are *not* demonstrating the leadership on this issue that is necessary.

Corporate culture

In some cases, this may be because the council's corporate leadership and elected members remain unconvinced as to the local benefits of action. The increasing national policy focus on the issue, together with improved understanding of local climate change impacts, may help to remove this barrier.

Another issue can be that climate change is still seen as an environmental issue, rather than something that has a fundamental impact on local wellbeing in the broadest sense, and therefore something to be tackled by the council's environment department. If genuine leadership in this area is to be delivered, the council as a whole needs to act in a way that reduces council carbon emissions and encourages action in the wider community. Historically, local authorities have not worked across departmental boundaries, and overcoming this 'silo' mentality will be one of the greatest challenges here.

Linked to this is the need for expert knowledge on climate change mitigation and adaptation options. The rate of change of technologies in these areas is fast, and it is impossible for all officers concerned with elements of mitigation or adaptation to maintain their knowledge in this area. There are sources of external expertise available (see below) but councils also need to consider how best to develop an internal resource and to ensure that it is used across all departments.

Resources for climate change action

Often the biggest barrier to increased action is a perceived lack of resources. Councils need to accept that climate change poses serious risks to local wellbeing and therefore that significant resources should be devoted to mitigation and adaptation work. However, there are also a number of partnerships and sources of help that councils should also make good use of, which will ease the pressure on internal resources.

Partnerships

Local Strategic Partnerships have been mentioned previously. The example of Mid Bedfordshire's climate change conference is just one way that an engaged LSP group can bring additional expertise to the table. Equally, smaller district authorities that are brought together for the development of a county-wide LAA can build on this to work with one another on the development and delivery of carbon emissions reduction initiatives.

Some larger authorities already have good working partnerships with private sector organisations such as energy suppliers and energy-efficiency installers. It can be difficult for smaller authorities to engage the interest of these organisations: here again, working in a group, potentially within the Local Strategic Partnership framework can help.

Sources of further information and advice

There are a number of services available to help local authorities in their climate change work. For example, the Energy Saving Trust provides the 'Practical help' advisory service, which should be the first port of call for authorities looking for information on all aspects of sustainable energy in housing. Also, the Carbon Trust offers the Local Authority Carbon Management programme to help authorities take the lead by managing their own carbon emissions. In addition, DEFRA has recently announced funding of £4 million to help local authorities to fight climate change. This will fund a local authority best practice programme, aimed at sharing best practice and offering mentoring and training, and will be delivered through the new Regional Improvement and Efficiency Partnerships.[21]

Other resources that authorities may find useful include the Department for Business Enterprise and Regulatory Reform's Energy Measures Report,[22] the Nottingham Declaration Action Pack[23] and the Beacon Councils Sustainable Energy Toolkit.[24]

Now is the time to act

As this chapter has described, there is an expectation that local authorities will take a lead in national efforts to tackle climate change. Some authorities are already active in this area, and many others have made the initial commitment to action.

At the 2007 Local Government Association Climate Change Conference, the LGA's Chairman identified climate change as the most important long-term priority for local government and as a key test of its credibility and reputation. The evolving role of local authorities offers them the opportunity to take the lead on this issue, and to ensure that the country responds to climate change in such a way that the wellbeing of their local area increases as we move to a low-carbon world.

---

21 http://www.defra.gov.uk/news/2008/080311b.htm
22 www.dti.gov.uk/energy/environment/measures/page41270.html
23 www.energysavingtrust.org.uk/housingbuildings/localauthorities/NottinghamDeclaration/online_action_pack/
24 www.idea.gov.uk/idk/core/page.do?pageId=5747988

The current policy framework allows local authorities to choose whether or not they focus on climate change mitigation and adaptation as key elements of local area improvement. It is vital that all authorities do choose this focus.

If they do, they will need to make far greater use of the power of wellbeing than they have to date, and to expand the influence and effectiveness of their partnership working. They should also demand the necessary supporting framework from central government, including some of the more radical measures mentioned earlier in this chapter.

If they do not, central government will have to take steps to force action, either by ensuring that the performance assessment framework offers sufficiently robust censure for inactive authorities or by introducing a duty to act on climate change.

It may seem that the 'deadline' for the 2050 emissions reduction target is quite some time away, but in terms of social change and large-scale investment in low-carbon technologies that are needed it is not. Locally led action is needed, and it is needed now.

References

AEA Energy and Environment (2007) *Analysis to Support Climate Change Indicators for Local Authorities*, London: Department for Environment, Food and Rural Affairs

Ashford (2006) *Local Development Framework Core Strategy Submission Document, Policy CS10: Sustainable Design and Construction*, Ashford Borough Council

Boardman, B. (2007) *Home Truths: A low-carbon strategy to reduce UK housing emissions by 80% by 2050*, London: Friends of the Earth, Co-operative Bank

Camden (2006) *Climate Change in Camden – A Joint Effort: Camden's climate change action plan 2006-2009*, London Borough of Camden

Camden (2007) *Camden Together: Camden's sustainable community strategy 2007-2012*, London Borough of Camden

CLG (2006) *Strong and Prosperous Communities: the Local Government White Paper*, Cm 6939, London: TSO

CLG (2006b) *Code for Sustainable Homes: a step change in sustainable home-building practice*, London: Department for Communities and Local Government

CLG (2007) *Planning Bill*, Bill 11, House of Commons, London: TSO

CLG (2007b) *Building a Greener Future: policy statement*, London: Department for Communities and Local Government

CLG (2008) *National Indicators for Local Authorities and Local Authority Partnerships: Handbook of Definitions. First release (February 2008). Annex 4: Local Economy and*

Environmental Sustainability, London: Department for Communities and Local Government

EST (2007) *Green Barometer: measuring environmental attitude*, London: Energy Saving Trust

HM Government (2007) *The New Performance Framework for Local Authorities and Local Authority Partnerships: single set of national indicators*, London: Department for Communities and Local Government

IPPR (2007) *Warm Words II: How the climate story is evolving and the lessons we can learn for encouraging public action*, London: Energy Saving Trust

LGA (2007) *A climate of change: final report of the LGA Climate Change Commission*, London: Local Government Association

Metcalf, G. and Greenhalgh, L. (2007) *A local climate impacts profile: LCLIP*, Oxford: UK Climate Impacts Programme

Uttlesford (2005) *Uttlesford District Council Local Development Framework: Supplementary planning document – home extensions*, adopted November 2005

Uttlesford (2007) *Uttlesford District Council Supplementary Planning Document on Energy Efficiency and Renewable Energy*, Consultation draft, August 2007

WAG (2007) *Local Vision: Preparing Community Strategies (consultation draft). Statutory Guidance from the Welsh Assembly Government on developing and delivering community strategies*, Cardiff: Welsh Assembly Government

CHAPTER 5:
Planning: central to delivery on climate change

Naomi Ludhe-Thompson and Hugh Ellis

Introduction

The land-use planning system remains the most sophisticated form of economic, social and environmental regulation ever introduced in the United Kingdom. Without the planning system, there would be no effective control of the development of land at the local, regional and national level. Spatial planning has a major contribution to make in meeting the climate challenge in several different ways. It can promote policies which expect the highest standards of resource use and energy efficiency, and substantially reduce the need for and propensity to travel by unsustainable modes (car and plane). It can make more effective use of land and sustainable transport modes through connecting work, schools, shops and healthcare facilities. It can promote an efficient supply of renewable and low-carbon energy from decentralised sources, and maximise the efficient use of resources while minimising waste. It can adapt to drought, flood risk and heat waves, build resilience into the man-made environment and increase its biodiversity. Finally, it has the opportunity to incentivise markets for new technologies that will help in mitigating climate change.

What are the key factors for success?

To deliver successful action on climate change through land-use planning, it is essential for there to be strong, clear policy, national fiscal measures and financial incentives, well thought-out timescales with appropriate technology, and collaborative working with developers. Behind all this there must be commitment at local level, by people who understand the issues of climate change and the need to take action. Strong policy can ensure that a common approach is taken to addressing climate change throughout the planning system, ensuring that a shared set of skills and knowledge is built up in planning and development control departments, and within the developer community.

National fiscal measures, such as Germany's *Feed in Tariff*, could help to create a situation where investment is possible by communities and individuals, in this case, in renewable (particularly solar) micro-technology. Financial incentives such as those related to energy efficiency and insulation, car sharing schemes and cheap public transport, are essential in changing consumer behaviour and market failure where possible.

Specific proposals and developments require close working with developers to deliver the kind of standards required both in the re-use or upgrade of existing buildings, and in new developments.

Local commitment is perhaps the key factor. The community, local government and local councillors need to agree that climate change must be addressed, and through the development of planning policies, reach consensus on strong, robust measures which may be challenging, but will deliver change on the ground.

Governing Sustainable Cities (Evans *et al.*, 2004) set out the key elements for developing capacity for sustainability within local government, based on extensive research in over 40 European cities. These elements included the ability to learn as an organisation, integrated working between departments, collaborative working, creative policy-making, facilitation and leadership, and public communication. In recent developments within the land-use planning system in the UK, there has been greater recognition for instance of learning from leading planning authorities and their successes in policy-making and implementation (CLG, 2008), and continuing calls for the planning department to be central to the local government function. Collaborative working has been enhanced with the development of Sustainable Community Strategies and Local Strategic Partnerships (LSPs). Creative policy-making has been very successful in changing the face of planning in the UK – Merton is the much-cited example – where a policy developed at local level, has been rolled out across the UK. Facilitation and leadership are skills that need to be further developed by planners and councillors to ensure success. Communication of the need to address climate change is essential to get the community on board, and to create a situation where participation in local plan-making brings together solutions and consensus. *Planning portal* and other web and electronic developments such as online consultations on planning authority web sites have helped in making large unwieldy documents more accessible through better navigation and accessibility features.

The land-use planning system: an overview

The fact that planning regulation places local decision-making by elected representatives at the heart of the process is one of its unique and most important aspects. There is also a powerful principled case for public participation in local planning decisions. This case is based on the importance of community empowerment and learning, encouraging active citizenship and delivering better informed decisions that themselves contribute to social and environmental justice and action on climate change.

Town and country planning developed from public health and housing policies. The growth in population and urbanisation developments in the nineteenth century led to problems which required government intervention. After a series of legislative measures, the 1947 Town and Country Planning Act '*...brought almost all development under control by making it subject to planning permission'*

(Cullingworth and Nadin, 2006). This meant the right to develop land was nationalised. Subsequent acts have changed the nature of planning, with recent governments undermining the system, and shifting the emphasis in development control back to market forces (Barker, 2007; House of Commons, 2007c) – but the planning system still survives.

European and UK legislation

European directives and laws directly influence national planning legislation, as all member states need to interpret and implement them. International agreements such as that on climate change also have a bearing on planning policy. There are also primary and secondary acts of parliament, key ones being the Town & Country Planning Act of 1990 and the 2004 Planning & Compulsory Purchase Act (England and Wales). The 2004 Act introduced many new provisions and inserted new sections into the 1990 Act. Essentially the acts taken together define the nature of development, the control of development and the plan-making system in England and Wales. The system operates in a similar way across the whole of the UK, but Scotland and Northern Ireland have different legislative provisions (Town and Country Planning (Scotland) Act, 1997, Planning etc. (Scotland) Act, 2006, Planning (Northern Ireland) Order, 1991). Northern Ireland in particular differs in that the Planning Service acts as the de facto planning authority, with six divisional planning offices. Local authorities in Northern Ireland only have a consultative role, which is very different to the decision-making role they have in England, Wales and Scotland.

Sustainable development duty

While the legislation does in general refer to the responsibility for planning to *'…contribute to the achievement of sustainable development'*, which is both unenforceable and unclear, the Planning Bill (House of Commons, 2007c) has introduced a new duty relating to local development documents in England to *'…include policies designed to secure that the development and use of land in the local planning authority's area contributes to the mitigation of, and adaptation to, climate change'*.

National planning policy

Development control policy and local plan policies are developed in the context of national planning policy. These are published in the form of planning policy guidance statements (PPSs) and minerals policy statements (MPSs) in England, and have legal status. Government circulars, government white papers and ministerial statements also carry weight. These are currently produced by the Department for Communities and Local Government, which holds the local planning function for England. In Wales, Planning Policy Wales (PPW) and Ministerial Interim Planning Policy Statements (MIPPS) are prepared by the Welsh Assembly Government. It also produces Technical Advice Notes (TANs). Scotland and Northern Ireland also develop their own separate planning policy.

Regional planning bodies and local planning authorities need to take into account these national planning policies when drawing up development plans and other documents and making decisions on planning applications. This is important in particular for climate change policy.

National frameworks

The planning system in Scotland has recently been overhauled with the 2006 Act , which has established a national planning framework for Scotland. Wales has had the Wales Spatial Plan since the 2004 Act, which is currently undergoing review. Northern Ireland has no framework plan, nor does England, although there have been calls for the Planning Bill to contain a national spatial framework for sustainable development for England.

Regional frameworks

There are however regional spatial strategies (RSS) in England (examples being Yorkshire & the Humber and the North East), which are blueprints for a region with a 15 year horizon, revised every five years. Many of them include important climate change policies, trajectories and targets. Alongside these lie the regional economic strategies (RES), regional sustainable development frameworks (or equivalent) and other regional strategies. An example is the emerging regional spatial strategy for the North West which sets out a number of policies for climate change and its implications. They include policy EM16 (Energy Conservation and Efficiency) amongst others, which states that local authorities must ensure that their approach to energy is based on minimising consumption and promoting maximum efficiency and minimum waste in all aspects of local planning, development and consumption.

In Wales, the RSS equivalent is the Wales Spatial Plan, which attempts to set out spatial sustainable development for the next 20 years. There have been criticisms of this plan, in its lack of integration of environmental, social and economic issues, and also in its evidence, in particular on the climate change impacts of both the plan as a whole, and the individual developments outlined. Scotland's National Planning Framework '...outlines a vision for Scotland's development to 2030. It builds on the government economic strategy and identifies key developments to support sustainable economic growth' (Scottish Government, 2004). Climate change is mentioned but there is no sense that the document is imbued with the understanding of the changes required to address the challenge.

Local plans

This regional tier influences local plans. Old style plans are being progressively replaced by local development frameworks in England, which are made up of statutory development plan documents and non-statutory supplementary planning documents. The local development plan system in Wales is equivalent to the local development framework in England, and is rather simpler, consisting of a single plan document which includes a core strategy, maps, policies and proposals. In Scotland, the development plans consist of two separate parts, the structure plan and the local plan. These set out how much development will take place, where it will take place, and where it will be protected from development. The development plans contain policies on use of land in an area, covering issues such as retail, transport, leisure, employment and conservation. The structure plan covers a wider area and takes a long-term view, while local plans usually cover smaller areas and become the basis for decision-making in that area. In Northern Ireland, plans for development vary between local plans, area plans, or subject plans.

Planning decision-making

The planning decision-making process is similar between England, Wales and Scotland, but differs in Northern Ireland because of the unique role of the single planning authority there. Planning policy is developed:

- Nationally, by Communities and Local Government (England), Welsh Assembly Government (Wales), Scottish Government (Scotland) and the Northern Ireland Planning Service (Northern Ireland).
- Regionally, by Regional Assemblies in England.
- Locally, by local planning authorities across England, Wales and Scotland.

The policies are contained within the local plan, are consulted upon, and examined in an inquiry. Regional spatial strategies undergo an 'examination in public'.

Some developments do not require permission – for instance the government in England has recently announced that certain types of micro-energy generation technology will be permitted by a general permitted development order, or a local development order, or it may not be classified as 'development'.

If the proposal requires permission, there should be pre-application discussions to ensure the proposal is in line with policy. The application is then submitted and there is an opportunity to comment, and depending on the nature of the application, other procedures and statutory consultees. The planning officers then prepare a report, taking into account national, regional and local planning policies, and make recommendations for refusal or approval by the planning committee. Minor decisions can be delegated to the planning officer. Refusal requires clear explanation in relation to planning policies. Appeals on refusal can be made, and then an independent inspector (from the Planning Inspectorate in England and Wales) runs an inquiry – either through written representations, an informal hearing, or a public local inquiry, and makes a decision. Some decisions can be taken by the relevant Minister rather than inspector if the application is of a certain size or significance. A challenge can be made to this decision via judicial review, on procedural rather than policy issues.

Local government structure

The local government structure differs between England, Wales, Scotland and Northern Ireland. Wales and Scotland have no two-tier councils, and therefore have a single local plan, and a 'unitary authority'. In England, two-tier areas have additional waste and transport plans sitting at county council level, while the district council holds the local development framework. Northern Ireland has unitary authorities, but no plan-making or development control powers – they are only consultative. Changes have been announced recently, with 11 new local government districts proposed, with some devolution of local planning.

National planning policy on climate change

It is clear that national planning policy must address the challenge of climate change comprehensively. Developments in renewable energy technology, building technology and pioneering local authorities have driven changes in national planning policy, but while climate change has been referred to and described in policies since the early 1990s (Climate Change Convention, 1992), the roadmap for planning authorities has been patchy. The *UK Sustainable Development Strategy* (1999; 2005) identified the planning system as the key lever in helping to meet climate change emissions reductions targets. In order for climate change action to be taken through planning, it is necessary to spell out the different policies which address climate change that can be framed at local level, and also the kind of development and adaptation which should be the result of these policies. Both mitigation and adaptation policies should be integral to all local plans across the UK.

The situation is changing rapidly with developments in legislation and policy across the UK. But there is no UK-wide agreement on action for local planning authorities which builds on the *UK Sustainable Development Strategy*, nor does the Climate Change Bill (House of Commons, 2007a) contain duties on planning authorities to reduce carbon emissions.

England

Developments in England have been both in legislation and in policy. The Energy Bill is largely concerned with energy infrastructure, and includes the renewable energy obligation to encourage investment in renewable energy technologies (House of Commons, 2008a). The Planning and Energy Bill (House of Commons, 2008b) is specifically concerned with local planning authorities to set requirements for energy use and energy efficiency in local plans. The bill is very short and states that:

> '(1) A local planning authority in England may in their development plan documents, and a local planning authority in Wales may in their local development plan, include policies imposing reasonable requirements for –
> (a) a proportion of energy used in development in their area to be energy from renewable sources in the locality of the development;

(b) a proportion of energy used in development in their area to be low
    carbon energy from sources in the locality of the development;

(c) development in their area to comply with energy efficiency standards that
    exceed the energy requirements of building regulations.'

This is essentially the transfer into legislation of a climate change planning policy, driven by the iconic 'Merton Rule' (Merton, 2003), which required a 10 per cent target for the use of onsite renewable energy to reduce annual carbon dioxide emissions for all new major developments. Recent government policy on climate change for England has been published as an Annex to Planning Policy Statement 1: *Sustainable Development* (CLG, 2007a). This policy supplement sets out how planning should contribute to reducing emissions and it is in addition to Planning Policy Statement 22: *Renewable Energy* (ODPM, 2004), and other statements on flood risk, housing and sustainable design.

The PPS1 Supplement includes new requirements for local planning authorities to ensure that tackling climate change becomes a primary concern for planning policy development and decision-making, stating that there is *'...an urgent need for action on climate change'* (CLG, 2007a). The recently published draft Planning Policy Statement 4: *Economic Development* (CLG, 2007b) seems to mitigate its impact somewhat, stating that *'Regional planning bodies and local planning authorities should plan to encourage economic growth',* which seems in direct contravention to the idea of sustainable development.

PPS: *Planning and Climate Change* (CLG, 2007a) also lists the issues that local planning authorities should take into account including:

- drive the delivery of including renewable and low-carbon energy;
- using place shaping to encourage viable resource use, energy efficiency and reductions in emissions;
- reduce the need to travel alongside growth;
- consider social issues when developing places and ensuring they are resilient to climate change;
- conserve and enhance biodiversity; and
- respond to the needs of communities and businesses.

Essentially, climate change must become a consideration in all policy-making, spatial plans, and decision-making on planning applications. Local authorities are responding to this positively, for example as Torbay's planning department stated: *'The Council's emerging Core Strategy will be promoting a sustainable planning framework that fosters energy efficient and carbon neutral development in the context of this recent Climate Change Planning Policy Statement'* (Torbay, 2008).

It recommends that targets and trajectories are used in the regional spatial strategies where policies can deliver carbon dioxide emissions reductions. Targets for the percentage of renewable energy generation on new development should be firmly evidence-based, with the possibility to raise development or site area targets where renewable and low-carbon resources afford greater potential. The requirements for the type of development which would be approved under the new PPS1 Annex are extensive, comprising energy supply, design, orientation (for passive solar gain), multi-functional greenspaces for biodiversity and people both public and private, sustainable urban drainage systems (SUDS), sustainable waste management and sustainable transport. It also confirms the central role of planning in helping to achieve the government's aim of zero-carbon development for all new homes from 2016. Following a report by the UK Green Building Council, commissioned by the Department of Communities and Local Government (CLG), a similar approach is likely to apply for new non-domestic buildings and development in order to reduce carbon dioxide emissions from these sources.

Wales

The Planning and Energy Bill (House of Commons, 2008b) also requires Welsh planning authorities to develop policies around energy use and efficiency. Planning Policy Wales (WAG, 2002) is due for an update on climate policy in the form of a Ministerial Interim Planning Policy Statement shortly. Wales has been dramatically pushing forward the climate change agenda over the last year – publishing its energy route map (WAG, 2008) with its ambition for Wales to be self-sufficient in renewable energy. The draft planning policy on climate change will require new developments to reduce their predicted carbon dioxide emissions by a minimum of 10 per cent, as well as outlining considerations in travel patterns, design, location, density, and access to renewable energy resources. The Strategic Environmental Assessment (SEA) is seen as the way to assess the total impacts of the plan and policies on climate change, and to reduce carbon dioxide emissions.

Scotland

The Scottish Climate Change Bill proposal (Scottish Government, 2008) contains a suggestion of possible supporting measures (to the overall carbon budget targets and monitoring) which could create a requirement for public sector bodies (local authorities and large public bodies) to take specific action on climate change, produce guidance, and make regular reports on the actions they are taking to address climate change. There is no specific national planning policy on climate change in Scotland, but there is policy on renewable energy, transport and flooding.

Northern Ireland

Northern Ireland's general planning policy statement has not been recently revised to reflect the policy approaches for tackling climate change seen in England and Wales. Sustainable development is a concern, but the detailed assessment of the carbon dioxide impacts of developments, and the measures required to reduce those impacts are not present.

The picture across the UK can therefore be viewed as fragmented, with some core sustainable development policies, but with a wide variation in the level of detail and prescription as regards the specific approaches to deal with climate change through different sectors.

How planning can influence sectors

Planning policy, supplementary planning guidance (SPD), and the proposals map are the key elements of most local plans in the UK, although they may be titled differently. The Strategic Environmental Assessment, and/or Sustainability Appraisal are part of the climate change toolkit as they should ensure that climate change impacts and causes are at least assessed.

A suggested policy approach to proofing all policies in order to mitigate climate change impacts in local plans is as follows:

1. Establish robust baseline data on CO_2 emissions.
2. Analyse in detail the potential impacts of the new policy options when assessed against these baseline conditions.
3. Establish transparent CO_2 reduction trajectories.
4. Review the suggested policy to deliver on the reduction trajectories (this should emphasise delivery and consider all mechanisms).
5. Monitor and review the effect of local plan policies on carbon emissions reductions and increases as far as possible.

Site allocation is important in considering the impacts of climate change, namely flooding and sea level rise, in considering access to resources, and location-specific issues. The sustainability appraisal of the plan should ensure these issues are appropriately considered and acted upon.

Overarching climate change policies have not been developed in a stand-alone way by local plans in the UK. Some local authorities have committed to action on climate change in their core strategy or their objectives (for example, the *London Plan*), but generally, the option has been to work on a sector to try and reduce emissions. Those planning policies which address climate change can be grouped into sectors of influence, covering energy, housing, transport, access and design, waste and resource use.

Energy planning policy

PPS 22: *Renewable Energy* (ODPM, 2004) sets out England's commitment to the development of renewable energy technologies. Wales has Technical Advice Note 8 on Renewable Energy (WAG, 2005), which includes 'strategic search areas' – general areas for the development of wind farm schemes. Scotland also has policy on renewable energy and Northern Ireland has a draft policy, both of which encourage the development of these technologies.

Since the introduction of national planning policies, local authorities have developed their own energy planning policies. The Greater London Authority in particular has developed an ambitious target in the Energy Strategy adopted in the 2004 plan:

> 'The Energy Strategy sets targets for the reduction of carbon dioxide emissions by 20 per cent relative to the 1990 level by 2010 as the crucial first step on a long-term path to a 60 per cent reduction from the 2000 level by 2050' (GLA, 2007).

The *London Plan* functions in a similar way to a regional spatial strategy with regard to local planning authorities in London, and the Mayor has the power to direct local planning authorities to refuse permission. The *London Plan* policy which had an impact was framed as follows:

> 'Policy 4A.9 Providing for renewable energy:
> The Mayor will and boroughs should require major developments to show how the development would generate a proportion of the site's electricity or heat needs from renewables, wherever feasible.'

This policy's effect was assessed by London South Bank University (LSBU, 2007) in a recent report. They stated that:

> 'In general, the Mayor's policies have been highly successful in reducing expected energy consumption and CO2 emissions in new developments. There has been strong upward trend in cumulative CO2 savings, with overall savings of approximately 135,528 tonnes CO2/year since the introduction of the London Plan, representing around a 26% saving of CO2.'

They particularly note the importance of the learning curve, whereby developers and planning officers struggled with the implementation of the policy in the first few years, but through experience, both developed the knowledge and skills required to deliver the requirements of the policy. Building on this success, the revised *London Plan* has now increased its target further:

> 'Policy 4A.1 Tackling climate change:
> The Mayor will, and boroughs should, in their DPDs require developments to make the fullest contribution to the mitigation of and adaptation to climate change and to minimise emissions of carbon dioxide. The following hierarchy will be used to assess applications:
> * using less energy, in particular by adopting sustainable design and construction measures (Policy 4A.3)
> * supplying energy efficiently, in particular by prioritising decentralised energy generation (Policy 4A.6), and
> * using renewable energy (Policy 4A.7)' (GLA, 2007).

These policies are contained within a section framed as the measures to deal with mitigation and adaptation to climate change. In particular this contributes to Objective 6 of the London Plan which states its ambition to make London a world leader in addressing climate change.

Housing planning policy and building regulations

The Planning Policy Statement on Housing, the new Housing Planning and Delivery Grant and the new Code for Sustainable Homes (discussed elsewhere) in England demonstrate the government's difficulty with housing, planning and climate change. While the Code for Sustainable Homes sets out a route map for greening housing development, the grant is simply based on exceeding quantitative targets on housing delivery. Local planning authorities in England will have to grapple with both the drive in policy to deliver housing, and the drive to consider climate change when putting together their own policies in their Local Development Framework and in development control.

In Wales the situation is different, as the planning grant is not geared to housing targets. However, Wales' policy on climate change is out of date, and not as visionary as CLG's recent supplement to PPS1 (CLG, 2007a). Building regulations have not yet been devolved to Wales, while planning has, creating a rather confused situation between the Code for Sustainable Homes which doesn't mention Wales, but technically applies, and Planning Policy Statement 3: *Housing* (CLG, 2006a) which does not apply to Wales.

Scotland has published a consultation document on its housing policy in October 2007 (Scottish Government, 2007). It frames the delivery of new build (35,000 new homes a year) within the need to consider, for instance, water, flood and other sustainable development issues, and setting higher environmental standards for buildings.

Some examples of the development of housing policy have been included in community strategies, such as Crewe and Nantwich Borough Council's *Sustainable Community Strategy 2006-2016* (2006), which states that the council will '...*encourage developers to use sustainable materials and technologies in future housing and commercial developments*' (Crewe and Nantwich, 2006). Although this is not a planning policy, it demonstrates integration between the aims of the community strategy and the local development framework.

Brighton and Hove City Council has been preparing a Sustainable Building Design SPD with targets for CO_2 reduction, low- or zero-carbon technologies and adherence to standards such as the Code for Sustainable Homes and EcoHomes. It also sets standards on other sustainable design issues such as SUDS, climate adaptation, energy efficiency, waste, sustainable materials, water reduction, etc. They hope to adopt the SPD later this year having revised the wording in line with the PPS. They are already working to ensure that the policies in the PPS1 are implemented in Brighton and Hove's plan, and will be working on incorporating it into the core strategy for adoption by 2010.

Sustainable design of developments

The planning system has a crucial role to play in considering overall development design issues, and should be setting out for developers a blueprint for design that is flexible enough to be applied in different areas of the council. Rural and urban areas will vary in terms of their design needs, as energy, water and transport resources will look very different.

Most examples of sustainable design are contained in supplementary planning documents – but Wrexham's *Unitary Development Plan* (2004) set out a comprehensive approach to designing developments in Policy GDP1:

> *'Policy GDP1 : Development Objectives:*
> *All new development should:*
> a) *Ensure that built development in its scale, design and layout, and in its use of materials and landscaping, accords with the character of the site and makes a positive contribution to the appearance of the nearby locality.*
> b) *Take account of personal and community safety and security in the design and layout of development and public/private spaces.*
> c) *Make the best use of design techniques, siting and orientation in order to conserve energy and water resources.*
> d) *Ensure safe and convenient pedestrian and vehicular access to and from development sites, both on site and in the nearby locality.*
> e) *Ensure that built development is located where it has convenient access to public transport facilities, and is well related to pedestrian and cycle routes wherever possible.'*

The policy goes on to include consideration of issues such as flooding, soil erosion, the avoidance of pollution, and the safeguarding of wildlife sites and habitats, and the creation of new sites where loss is unavoidable as part of the development. This comprehensive approach to development provides the outline for the developer of what detailed design they should achieve.

Urban areas face great constraints on certain sites, and planning authorities have responded to the challenges in different ways. Haringey Council (2003) has developed supplementary planning guidance on the need to assess and avoid contributing to air pollution through urban development, measuring the carbon emissions for a development in relation to baseline emissions and the projected increase or decrease (through predicted transport levels, construction, predicted energy use in the house).

Sustainable transport plans

Larger authorities such as Nottingham, Birmingham and Manchester have had successes in reversing the car-based urban transport plans of the 1970s. Nottingham pedestrianised its centre, as did Birmingham, despite public opposition in the former, and traders' opposition in the latter. Both are now public spaces which are thriving, and with a much reduced carbon footprint. Manchester's tram system has helped change the mobility patterns, but bus services and links are still problematic. The Local Transport Bill (House of Commons, 2007b) is expected to help reverse this situation and make services fit for purpose.

Bristol and Gloucestershire have been successful in developing car-sharing initiatives alongside new developments. Bristol's preferred options policy states:

'In 2026, we want Bristol to have...a pattern of development and urban design that promotes good health and well-being and provides good places and communities to live in. Bristol will have open space and green infrastructure, high quality healthcare, leisure, sport, culture and tourism facilities which are accessible by walking, cycling and public transport. This will help enable active lifestyles, improve quality of life and reduce pollution' (Bristol City Council, 2008).

This objective will be integrated with other policies on carbon emissions, green infrastructure, urban design and transport accessibility.

Barriers to tackling climate change

Despite positive impacts of planning, there are several barriers to addressing climate change through planning in the UK.

Culture change

Planning is not seen as a solution by government departments, with successive planning reforms serving to promote the interests of the private sector (Barker, 2006), rather than examining the contribution and role that the planning profession has made and should have in tackling climate change. Planners themselves suffer from a poor perception of their skills and role, and neglect within local planning authorities. Calls for change have been increasing, including the need to 'put planning first' (McNulty, 2003). The change in culture needs to go hand in hand with a much better quality of approach, but the starting point is to acknowledge the role that planning has to play, and based on evidence of where planners have been successful, to disseminate that as part of the change in culture. Efficiency, consistency, transparency and accountability must characterise the profession (ibid.) if it is to successfully engage people and the private sector in tackling climate change.

Councillors are also under pressure to start making decisions which address climate change.

The real impetus for change comes however from the urgency of the need to address climate change. It will become untenable for planning authorities to shirk their duties in helping their communities adapt to the future.

Knowledge and skills base

Addressing climate change requires different knowledge. Measuring the carbon impacts of certain policies requires an understanding of the causes and impacts of emissions, and the ways they can be reduced. It requires an understanding of policies which are effective in reducing emissions, and constantly updating knowledge of best practice and adapting this successfully. It also requires negotiation skills to persuade developers to build in the most carbon-efficient way that is fit for purpose, and puts people and the environment at the heart of the matter, rather than the developer's profit margins. Community engagement and involvement skills are essential, including knowledge of how to run transparent and representative processes. Communicating with and understanding community needs, fears and aspirations, and effectively incorporating them into the process of plan-making (policy development and decision-making) are vital to success.

The real situation is that planners, and councillors, lack the knowledge and the skills they need to address climate change. Large metropolitan authorities are often the best qualified, with better pay, opportunities and dynamism characterising departments. Small rural districts suffer in recruiting staff, and in the lack of professional development, enthusiasm or ability in existing staff.

The Royal Town Planning Institute runs numerous training courses and conferences to enhance the planning profession's knowledge and skills base, and the Improvement and Development Agency and other government bodies in England, Wales, Scotland and Northern Ireland are providing opportunities for planners to develop their knowledge and skills base further.

Evidence base

Planning policy and decisions should not be made in a vacuum. They should be grounded in robust evidence, from a variety of sources. Collecting this evidence is often poor, although the Strategic Environmental Assessment is helping to develop better collection of data on issues such as climate, air, water, soil and so on. Climate change policy must be firmly based in evidence and what is achievable, while still being ambitious and creative. Carbon emissions for a development, for instance, include the construction materials, energy use in building, energy use in living in the building, and transport associated with the building. The Code for Sustainable Homes and BREEAM provide guides and methodologies for assessing the emissions. For policies, the assessment of the probable impact of a land-use planning policy on carbon emissions requires careful accounting, but starts from the principle that emissions for vehicle use and therefore types of journeys are measurable, and energy use is also measurable. Reducing carbon emissions overall as part of the plan's core strategy must therefore be able to gather together the evidence on reducing emissions from waste, from journeys, from energy use, from industry in the area, from business, from retail; and the increases in emissions which result in the change of land use or other impacts such as soil changes or biodiversity changes/loss.

Community engagement

The development of the local plan provides essential opportunities for people to be involved. By being involved in the development of the plan, it is hoped that communities will increase their understanding of climate change, and what needs to happen in order to address climate change. However, there are barriers in the language of planning, the methods of involvement, and the complexity of the issues (Townsend and Tully, 2007). These barriers must be overcome to develop capacity in civil society to tackle climate change.

A positive future

The land-use planning system in the UK provides huge opportunities for the development of a low-carbon Britain and low-carbon communities. Nationally, priorities can be set through integrated national planning policy. This means that every piece of national policy should be assessed against the criterion of delivering carbon emissions reductions and should positively contribute to sustainable development. Locally, priorities can be set which translate national policy into a local plan and supplementary planning guidance. The local plan should contain core policies which address sectoral issues such as housing development and business development, as well as infrastructure development such as transport, power, heating, water and waste water services. In terms of the physical environment, the key issues are flooding, urban heat, surface drainage water and biodiversity. All these policies need to understand their spatial and physical interaction to truly work.

Increasing numbers of local authorities have strong climate change policies as part of their core strategies, whether in their Local Development Frameworks (LDFs) in England or Local Development Plans (LDPs) in Wales, or local plans in Scotland and Northern Ireland.

The vision of a sustainable community delivered through planning

Freiburg, located in the Black Forest region of Southern Germany, is an example of a city which has successfully managed to grow and develop, but has used its planning system to ensure that new developments are sustainable, and to foster a more sustainable lifestyle. It is also a movement, where the planners' vision of a low-carbon place is actively welcomed by the community itself.

Freiburg has two areas of development on the outskirts of the city, Rieselfeld, and Vauban. In these new developments, social needs are met through varied sizes of homes, mainly in shared flats, with shared gardens and children's playgrounds. Each grouping has nursery childcare facilities provided. Central to both new developments is the community centre, which in the larger development, Rieselfeld, consists of a unique shared church, and secondary school, and an impressive green space. New sports and leisure facilities have been developed, but initial commitments to green space remain, with a green roof built over the extra indoor sports courts. This is important because it delivers environmental goods to the community as well – by reducing travel to these facilities it reduces climate change emissions. These social functions also deliver jobs, both of themselves, and in the services and products they require and generate a market for – at a local, sustainable economic level. Again this helps to address climate change, by reducing travel and related energy use.

In terms of the environment, natural streams are maintained throughout the development, and a buffer zone is maintained to the south of the development, and is protected as a nature reserve. As the buildings are built around large irregular courts, there is minimum surfacing, as these courts are usually green spaces, consisting of the gardens or playgrounds with pathways. The buildings themselves are highly energy efficient, saving on heating costs, and solar panels are widely used for supplying energy.

Transport is in the form of a dedicated tramline from the centre of the city, which interlinks with the bus services in the centre of the new development. Both new developments have had dedicated tramlines built. Despite a growth in population, Freiburg has actually managed to reduce car use, with a phenomenal rise in cycling and public transport journeys. The smaller of the two developments, Vauban, is planned as a car-free development, with no dedicated car-parking spaces, and car use actively discouraged through very high charges on the parking spaces that exist. (FoE, 2007)

Make or break

This is a crucial stage in the history of planning in the UK. The Planning Act 2008 may drive the UK towards a high carbon infrastructure, or if climate change is recognised as an overwhelming consideration in decision-making it could be an opportunity to move towards a low-carbon Britain. Vital infrastructure developments will now be approved by the Infrastructure Planning Commission rather than local planning authorities (including roads, energy developments, waste facilities), and therefore the local plan will not have as strong a statutory role. If the plan-led system is allowed to fade away through lack of support or understanding, local government's ability to act on climate change and be a true place shaping force will be severely limited. The democratic deficit in this country will increase and participation in decision-making will be ever more reduced.

National policy statements on major infrastructure must drive forward action on climate change, and work with the government's stated commitment to climate change policy in its supplement to Planning Policy Statement 1.

Conclusion

Governments in the UK must stop pulling local planning authorities in different directions by asking them to address climate change, while giving far greater power to developers. The outcome of having a developer-led system rather than a robust plan-led system is that the public interest will be lost. Developers are not able to consider the strategic requirements of an area in terms of climate change, and delivering social and environmental justice. Our local authorities and our local government should and must have a strong vision of a just, low-carbon community which developers must work within.

References and further reading

Boardman, B. (2007) *Home Truths: A low-carbon strategy to reduce UK housing emissions by 80% by 2050*, London: Friends of the Earth

Brighton and Hove Council (2008) *Draft Sustainable Building Design SPD*, Brighton: Planning Department

Bristol Council (2008) *Preferred options – Local Development Framework*, Bristol City Council

CLG (2006a) *Planning Policy Statement 3: Housing,* Communities and Local Government

CLG (2006b) *Planning Policy Statement 25: Development and Flood Risk*, Communities and Local Government

CLG (2006c) *Code for Sustainable Homes: A step change in sustainable home-building practice*, Communities and Local Government

CLG (2007a) *Planning Policy Statement 1: Sustainable Development, Climate Change Supplement*, Communities and Local Government

CLG (2007b) *Consultation Paper on new Planning Policy Statement 4*, Communities and Local Government

CLG (2008) *Best Practice Guidance on Climate Change*, Communities and Local Government

Crewe and Nantwich Borough Council (2006) *Sustainable Community Strategy 2006-2016*, Crewe and Nantwich Borough Council

Cullingworth, B. and Nadin, V. (2006) *Town and Country Planning in the UK*, Routledge

Ellis, H. (2007) *Draft Planning Policy Statement on Climate Change*, Friends of the Earth and Town and Country Planning Association

ERM and Faber Maunsell (2008) *Working Draft of Practice Guidance to support the Planning Policy Statement: Planning and Climate Change*, Environmental Resources Management, Faber Maunsell. Available from: http://www.erm.com/practiceguideance

Evans, B., Theobold, K., Joas, M. and Sundback, S. (2004) *Governing Sustainable cities*, Earthscan

FoE (2007) *Report from study visit undertaken to Freiburg im Breisgau, Germany, in December 2007*, Friends of the Earth

GLA (2007) *The London Plan (consolidated with alterations since 2004) – the Mayor's Spatial Development Strategy*, GLA

Haringey Council (2003) *Supplementary Planning Guidance Air Quality*, Haringey Council

House of Commons (2007a) *Climate Change Bill*, London: TSO

House of Commons (2007b) *Local Transport Bill*, London: TSO

House of Commons (2007c) *Planning Bill*, London: TSO

House of Commons (2008a) *Energy Bill*, London: TSO

House of Commons (2008b) *Planning and Energy Bill*, London: TSO

LSBU (2007) *Review of the impact of the energy policies in the London Plan on applications referred to the Mayor* (Phase 2), London South Bank University

McNulty, T. (2003) *Speech to the Local Government Association Conference*, Tony McNulty MP

Merton Borough Council (2004) *The Merton Rule*, Merton Borough Council

ODPM (2005) *Planning Policy Statement 1: Delivering Sustainable Development*, ODPM

ODPM (2004) *Planning Policy Statement 22: Renewable Energy*, ODPM

Scottish Government (2004) *National Planning Framework for Scotland*, Scottish Government

Scottish Government (2007) *Firm Foundations: The Future of Housing in Scotland*, Scottish Government

Scottish Government (2008) *Climate Change Bill consultation*, Scottish Government

Sustainable Development Unit (1999) *A better quality of life – strategy for sustainable development for the United Kingdom*, TSO

Sustainable Development Unit (2005) *Securing the Future – UK Government sustainable development strategy*, TSO

Torbay Council (2008) *Response to correspondence on PPS1 Climate Change Supplement*, Torbay Council

Townsend, A. and Tully, J. (2007) *Public Participation in the Revised Planning System*, Durham University

WAG (2002) *Planning Policy Wales*, Welsh Assembly Government

WAG (2006) *Planning Policy Wales Companion Guide*, Welsh Assembly Government

WAG (2008) *Energy Route Map*, Welsh Assembly Government

Wrexham Council (2004) *Unitary Development Plan*, Wrexham Council

CHAPTER 6:
Fuel poverty – social issues and sustainability

Peter Lehmann[1]

Introduction

With substantial energy price hikes over the past two years, more and more households find it difficult to pay their energy bills. A significant proportion find themselves in what is referred to as fuel poverty. A household is defined as being in fuel poverty if it has to spend more than ten per cent of its income to meet its heating and hot water requirements and any other energy needs.

This chapter examines the issue of fuel poverty in more detail by looking at the inter-relationship between social and environmental issues in housing. It describes the main fuel poverty programmes, especially those relating to energy efficiency, considers the relationship between fuel poverty and climate change policies, and critically reviews fuel poverty policies and the impact of and progress towards the government's statutory fuel poverty targets. This chapter focuses mainly on England. The devolved administrations have similar but not identical schemes (see BERR/ DEFRA, 2007).

Fuel poverty in the UK

A household's income, the price it pays for its energy and the quality of its housing, heating and other equipment will determine whether the household is in fuel poverty. The issues are broader than just those of fuel poverty – the quality of housing and equipment will have a major impact on the energy bills of low-income households generally. Fuel poverty is largely a problem of the existing housing stock, and is not an issue in newer housing because of the comparatively higher standards of energy efficiency. The largest concentration of fuel poor households is to be found in pre-1975 dwellings (BERR, 2005).

The government has, under the Warm Homes and Energy Conservation Act 2000, a statutory obligation to eradicate fuel poverty as far as is practical in the UK. Table 6.1 overleaf summarises the target dates.

---

1 Peter Lehmann was chair of the Fuel Poverty Advisory Group from February 2002 to March 2008.

Table 6.1: Statutory obligation to eradicate fuel poverty in the UK

	Vulnerable households	All households
England	2010	2016
Wales	2010	2018
Scotland	N/A	2016
Northern Ireland	2010	2016

Source: *The UK Fuel Poverty Strategy* (2007).

Households in fuel poverty in the UK

The numbers in fuel poverty fell for many years, but then they started rising from 2003 as Table 6.2 shows.

Table 6.2: Number of households in fuel poverty in England 1996-2006 (millions)

	1996	1998	2001	2003	2005	2006*
Total in fuel poverty	5.1	3.4	1.7	1.2	1.5	2.4
Vulnerable in fuel poverty	4.0	2.8	1.4	1.0	1.2	2.0
Social Housing	N/A	N/A	N/A	0.2	0.2	N/A
Private Housing	N/A	N/A	N/A	1.0	1.3	N/A

*2006 figures are estimated
Source: *The UK Fuel Poverty Strategy* (2007).

As can be seen from the table the reductions in fuel poverty between 1996 and 2003 were very substantial. Since 2003, however, there have been marked increases in the numbers in fuel poverty, so that in 2006 there were about 2.4 million households in England. The same is true in the rest of the UK as Tables 6.3 and 6.4 show.

Table 6.3: Number of households in fuel poverty in Scotland 1996-2005 (000s)

	1996	2002	2003-4	2004-5
Number of fuel poor households (>10% of income on fuel)	756	293	350	419
Number of extreme fuel poor (>20% of income on fuel)	182	71	112	119

Source: *The UK Fuel Poverty Strategy* (2007).

Table 6.4: Numbers of households in fuel poverty in Wales and Northern Ireland (000s)

	Wales	Northern Ireland
1998	380	N/A
2001	N/A	181
2004	130	126

Source: *The UK Fuel Poverty Strategy* (2007).

Reductions in the number of fuel poor households over the period 1996-2005 were driven by energy price decreases, as the combined product of world trends and the introduction of competition in the UK energy market. Household incomes also increased significantly during this period, especially with the introduction of Pension Credit and Tax Credits. Furthermore, the programmes for improving the energy efficiency of low-income households also made an important contribution.

The government, in its 2007 annual report on fuel poverty (BERR/DEFRA, 2007), estimated that – of the reduction in overall fuel poverty between 1996 and 2005 – three-quarters was due to income improvements and around a fifth to energy-efficiency improvements, with the rest resulting from price reductions. The subsequent sharp and disappointing increases in the numbers of households in fuel poverty were primarily the result of energy price hikes since 2003.

Although the energy-efficiency programmes have reduced the impact of rising prices, they have not been able to prevent this sharp rise.

Impact of reducing fuel poverty

A reduction in and elimination of fuel poverty, especially through improved energy efficiency, brings a range of benefits and helps with a series of government objectives. There will obviously be considerable similarities between the impacts of measures to reduce fuel poverty and others to improve housing more generally. Reducing fuel poverty will help to:

- enable people to keep warm;
- increase an individual's disposable income;
- reduce CO_2 emissions;
- reduce excess winter mortality – in part related to cold homes;
- reduce cold-related illnesses;
- enable more elderly people to live independently;
- make it possible to release people from hospital to their homes more quickly;
- reduce costs to the National Health Service; and
- reduce health inequalities.

Main fuel poverty programmes

There are very sizeable schemes that provide energy-efficiency measures free of charge for low-income households and these measures can, in many cases, significantly reduce fuel bills. The schemes also provide benefit entitlement checks to maximise an individual's benefit uptake.

Warm Front (England and Wales)

Under Warm Front, eligible customers receive energy-efficiency measures, such as insulation, draught-proofing and energy-efficient light bulbs free of charge. Those eligible are households with children and pensioners on means-tested or disability benefits and tax credits in private sector housing. Further measures include central heating installation. The scheme does not extend to the social housing sector. Current expenditure (2007/08) amounts to about £300-£350 million. Unfortunately this is being cut to about £270 million per annum over the next three years, but nevertheless remains a sizeable programme.

The Energy Efficiency Commitment (Great Britain)

The Energy Efficiency Commitment (EEC) is an obligation on suppliers of electricity and gas to secure energy savings from their domestic customer base, largely via insulation and efficient white goods. Half the benefits of the scheme have to go to the 'priority group' (i.e. those receiving means-tested benefits, tax credits or disability benefits). Many of the measures are provided free of charge to the priority group. The scheme covers both the private and social housing sectors. The scheme is a sizeable one with combined expenditure by the companies in the region of £200 million per annum on the priority group from April 2005.

There are six major companies involved – British Gas, EDF Energy (formerly London and South West Electricity), RWEnpower (formerly National Power), Eon UK (formerly Powergen), Scottish & Southern, and Scottish Power.

The scheme has been expanded (doubled in size and somewhat changed) and operates under a new name – Carbon Emissions Reduction Target (CERT) – since April 2008. Expenditure on a rather wider priority group is likely to be £400 million per annum.

Decent Homes Standard in the social sector

The drive to bring homes in the social sector up to the Decent Homes Standard has, of course, resulted in major programmes. A very high proportion of homes failing the Decent Homes Standard did so on thermal comfort (i.e. energy efficiency) grounds. Loft, cavity wall insulation and improved heating programmes are the main measures to meet the thermal comfort criteria. Insulation programmes under the Decent Homes Standard have been running at around £50 million per annum and central heating (mainly replacement of older systems) at £250-£300 million per annum. A significant part of the improvements in the social sector has been funded under EEC by energy suppliers working in partnership with local authorities.

However, the Decent Homes Standard is not a very demanding standard in terms of energy efficiency and over half of fuel poor households in the social sector were living in homes meeting the Decent Homes Standard. This is high, although the incidence of fuel poverty is nearly three times as great in homes which are not decent (this is discussed below in the section on policy issues).

Supplementary programmes

There are a number of other housing and heating-related programmes to reduce fuel poverty. These often draw in part on Warm Front and EEC funds.

Gas network extensions

In spite of the sharp gas price increases of recent years, gas central heating is generally, amongst conventional heating methods, the cheapest form of heating. Somewhat over ten per cent of households are outside gas supply areas and the incidence of fuel poverty is two to three times higher amongst them compared to other households. There are about 4,500 clusters of communities within 2 kilometres of the gas network but not connected to it. The Office of Gas and Electricity Markets (Ofgem) is likely, from April 2008, to introduce measures as part of the 'gas distribution price control' to facilitate the extension of the gas network to a number of relatively 'deprived' communities within a short distance of a gas main.

Household renewables

For 'deprived' communities further afield, the Design and Demonstration Unit of the Department of Business, Enterprise and Regulatory Reform (BERR) is working with some of the Regional Development Agencies on programmes of household renewables, especially heat pumps. The programme is in operation in three regions/devolved administrations, namely the North East, Yorkshire & The Humber and Wales, with five more planned for early 2008. Most of the projects are air-sourced heat pumps and involve 50-100 homes each.

Warm Zones and area-based approaches

Finally, the Warm Zones programme tackles fuel poverty on an area basis. In the initial Warm Zones, local authorities, especially in the North East, worked in partnership with the National Grid to 'blitz' an area, i.e. the whole, or more usually, part of a local authority area. Warm Zone employees would go from door to door, assessing the eligibility of households for the fuel poverty programmes, referring them where they were eligible and giving further advice where they were not.

There was also a great deal of local publicity and in many areas there were checks to ensure that households were receiving all the benefits to which they were entitled.

One of the most far reaching Warm Zones has been in Kirklees where every home which is suitable for loft and cavity wall insulation receives it free of charge. In addition, there are free low-energy light bulbs for all, free improvements of heating systems for certain householders and lower prices for renewable technologies for able-to-pay customers. Other help is also provided in the form of benefit checks, fire safety checks and water saving advice. The programme is funded by Kirklees Council, a number of energy companies and the Regional Housing Boards as well as other local partners.

Other local schemes promoted by energy companies have provided a range of help through partnerships with charities such as RNIB, Gingerbread and the Family Welfare Association.

Some of the local programmes were very successful and the Department for Environment, Food and Rural Affairs (DEFRA) has now provided, under its Community Energy Efficiency Fund, a little over £6 million in 2007/8 in England to stimulate such approaches. A total of fifty schemes successfully bid for this money. These schemes are now being put in hand.

Regeneration schemes and private sector renewal

Fuel poverty in new or refurbished properties as part of regeneration schemes is unlikely to be an issue, since they will have to meet the relatively high energy-efficiency standards of current building regulations. However, regeneration schemes do provide an opportunity to secure economies of scale in improving the energy-efficiency of homes which are not being intensively refurbished.

Funding of private sector renewal programmes is, however, being cut in some regions, as a result of the drive to increase new house building.

Programmes on incomes and prices

Any programme to improve the incomes of low-income households will also help to tackle fuel poverty. Specifically, in the context of fuel poverty there have been programmes to improve the uptake of benefits. Following reports by the Eaga Partnership Charitable Trust (which runs the Warm Front programme) and the energy companies of low-income households not qualifying for one of their programmes, benefit checks started to take place. These programmes were extremely successful and an average increase in benefits of £1,000 per annum has been secured for a sizeable number of customers. This led to the decision to offer benefit entitlement checks to *all* households under Warm Front. Most of the energy companies also offer these checks. There are issues, notably that some customers who are eligible nevertheless do not claim, but overall the programmes have proved well worthwhile.

The numbers in fuel poverty have risen sharply as a result of energy price rises. These are largely a result of world forces. However, prices have, in the view of the Fuel Poverty Advisory Group (FPAG), risen more than necessary, especially for low-income customers. Companies' margins appear to have increased, charges for networks are higher than they need to be and, above all, customers on prepayment meters or paying by cash or cheque – who tend to have lower incomes – are now paying as much as £140 per annum (over 15 per cent) more than those who pay by direct debit or online. The energy supply companies, facilitated by Ofgem, have introduced Social Tariffs for some of their low-income customers. This is helpful, but is not nearly enough to offset some of the adverse trends described above, especially the widening price differential between different forms of payment.

Taken together the policies to improve the energy efficiency of homes and heating are very sizeable. In 2007/8 about £650 million is being spent on programmes such as Warm Front, EEC Priority Group and decent homes (thermal comfort element). This is likely to rise in 2008/9, as a result mainly of the expansion of EEC. Although (as discussed in more detail below) rather more is needed, these are large programmes.

One of the concrete results is that the average SAP rating[2] of the homes of low-income households was in 2005 nearly 50, compared to 48 for all households. This is a reversal of the historical situation. In 1991 the energy efficiency of the houses of low-income groups was below that of all households, so that over a period of years, energy efficiency has increased more rapidly in the homes of low-income households than of others.

The relatively high SAP rating for low-income groups is partly due to the fact that social housing historically has had relatively high SAP scores. But it is also a result of the various fuel poverty programmes. There is still a huge amount to do, but this is nevertheless an impressive achievement. In a very few cases conditions for low-income households are even as good as those of people more financially secure.

Interactions of climate change and fuel poverty policies

There are considerable synergies between climate change and fuel poverty policies, but also some tensions. In order to combat climate change, the energy efficiency of the housing stock will need to be increased and carbon emissions reduced (see Chapter 10). This is true regardless of the socio-economic profile of the occupants. Hence, a wide range of energy efficiency and carbon reduction programmes are

---

2   SAP – the Standard Assessment Procedure for is a methodology to calculate the energy rating of a dwelling. SAP ratings are scored on a scale from 1 to 100 where 1 is the worst and 100 indicates the highest possible energy-efficiency rating.

needed. Sufficiently large programmes will help to reduce unit costs of micro renewables.

A general misconception is that a reduction in fuel poverty will increase emissions, as households are using more energy to keep warm. However, because significant improvements in energy efficiency can be made in many homes, households can have higher heating standards, lower energy bills and emit less carbon. Thus, in 2006/7 the Warm Front programme secured an average improvement in SAP ratings per household from 40 to 56, a fall in CO_2 emissions from 6.97 MtC per annum per household to 6.16 MtC and the potential to save an average £190 per annum for each household in energy running costs.

The main tensions arise on energy prices. The price of energy has risen and will continue to rise as a result of measures to reduce emissions. This is due to the fact that energy from low-carbon sources is more expensive than 'conventional' energy and incentives have to be provided for the take up of low- or zero-carbon generated energy. This underlines the importance of using the most cost-effective ways of reducing emissions, which has not been achieved so far in some of the important UK carbon reduction programmes. It also highlights a political point about the interaction between these policies. We are likely to be inhibited from pursuing climate change policies as vigorously as we might otherwise do, if there is continued concern about fuel poverty and low-income households in energy-inefficient homes.

Whilst there are both synergies and tensions, these can and will need to be managed in wealthy economies. If progress is not made on fuel poverty there is a risk that this will hamper effective climate change policies.

Policy issues

Resources

FPAG, in its 2006 Annual Report, concluded that programmes of about £1 billion per annum would be needed from 2008 to 2016 (along with other measures) in order to eradicate fuel poverty by 2016. The key capital measures are outlined in Table 6.5:

Table 6.5: Capital measures to tackle fuel poverty

Measure	Cost £ million
Mainstream insulation	260
Central heating (CH)	440
Non-gas areas (gas CH or renewables)	900
Other (gas CH replacement, solid walls insulation, solar thermal)	3,900
Total	5,500

Source: FPAG (2007).

These are purely the cost of the measures themselves. When account is taken of administration and marketing costs (since it is not known exactly who and where the fuel poor are), the estimated total cost is about £9 billion, or £1 billion per annum. This is a sizeable programme and some way above the level of the programmes for the next three years.

Decent Homes Standard

The decent homes programme in the social sector has been an ambitious and far reaching programme which is largely on target. Unfortunately the standards in terms of thermal comfort are very low (Box 6.1).

Box 6.1: Decent Homes Standard – thermal comfort standards

- Minimum of only 50 mm for dwellings with gas or oil central heating.
- Loft *or* cavity wall insulation, but *not both* if there is a gas or oil central heating system.
- Dwelling can meet Decent Homes Standard even if it has a very inefficient heating system, i.e. electric storage heaters, LPG, oil or coal central heating.
- A dwelling is counted as having central heating if two or more rooms are heated centrally.

Source: CLG (2006).

So, it is not surprising that, as noted, some households in homes meeting the standard are still in fuel poverty. Many social landlords are hoping to go beyond the strict Decent Homes Standard in their programmes. According to CLG research, social landlords are expecting 90 per cent of homes with gas or oil central heating to have both loft and cavity wall insulation. They further anticipate that 85 per cent of lofts (which can be insulated) will have 200 mm of loft insulation. Even if this is rather optimistic, it is clearly helpful, but it is still important both from a fuel poverty and a climate change point of view that the standard is raised.

FPAG recommends the standard to be raised after 2010 and that in the interim period the new higher standard should be used where work is in any case being carried out on a property, rather than having to carry out additional work to account for the higher standard in the near future. FPAG's proposal is that all social housing should achieve a SAP rating of at least 65 by 2016.

In this context it is worth noting that private sector properties under Warm Front already have to achieve a SAP rating of 65 wherever practical. It is hard to understand why social sector standards should be lower than those in the private sector when, in both cases, public funds are being used.

Local authorities

Local authorities clearly have a very important role to play in combating fuel poverty via a number of measures, namely:

- the Decent Homes Standard;
- the Home Health and Safety Rating System and measures to improve standards in Houses in Multiple Occupation (see below);
- Energy Efficiency Grants, especially the co-ordination of any local grants within the national schemes such as Warm Front and EEC;
- securing referrals to the fuel poverty schemes directly and through local partnerships; and
- benefits advice and take-up campaigns.

In the past, the fuel poverty issue was not in any way reflected in the local government performance measurement framework. This is now changing and both a fuel poverty and a climate change indicator have been included in the new 198 local government performance indicators. Whilst this is a step forward, there are some concerns about the precise indicator currently being proposed with regard to fuel poverty, i.e. the number of people receiving income-based benefits who live in homes with a very low energy-efficiency rating. This is very narrow as it only focuses on those homes with the very worst energy efficiency and there is no incentive to improve the homes of others in fuel poverty with a poor energy-efficiency rating. The climate change indicator (per capita reduction in CO_2 emissions in the local authority area) will also be helpful in tackling fuel poverty.

The next step will be to secure the inclusion of fuel poverty in Local Area Agreements (LAAs). This has already been done by a number of local authorities, especially during the first round in early 2006. One of the attractions of including fuel poverty in the LAA is that a great deal of progress can be made by bringing funds from the sizeable fuel poverty programmes into the local authority area.

Local authority access to information

The costs for local authorities of finding customers eligible for the fuel poverty schemes are higher than they need to be, because of the obstacles to sharing data. Local authorities need access to the information that will enable them to locate those who could benefit from the fuel poverty programmes and who might also usefully receive advice on taking up welfare benefits. FPAG has been strongly urging CLG to allow access for local authorities to the energy-efficiency data of the Energy Performance Certificates, so that they can effectively distribute information on the help available.

This is clearly important, both for the reduction of emissions and for tackling fuel poverty. CLG is considering this, but has so far not been willing to agree to it, which is disappointing.

Due to legal uncertainties, some local authorities are even unable to access the data on council tax and housing benefit recipients for the purpose of targeting the fuel poverty programmes *within their own authority*. The same arguments are taking place in a range of different local authorities with answers differing from authority to authority. This is a huge waste of time and the situation can only be described as chaotic. Thus, FPAG is urging government to clarify the legal situation once and for all.

The Housing Health and Safety Rating System (HHSRS) and Houses in Multiple Occupation (HMO)

The Housing Act 2004 brought in two measures which help in tackling fuel poverty in the private rented sector: mandatory licensing of HMOs, and replacement of the test of fitness for human habitation by the HHSRS. It is now important that this legislation is implemented effectively and that the necessary resources are made available. The legislation contains some energy-related provisions, specifically if the landlord refuses an offer of a free Warm Front or EEC priority group measure. This provides a *prima facie* case for an HHSRS inspection. A number of local authorities are using this provision and it is hoped that others will actively use it, as it appears to provide a win-win solution for both tenants and landlords – the energy efficiency of the home is increased, tenants' fuel bills are reduced and/or they can keep warmer and the measures are wholly, or largely, funded through the fuel poverty programmes. The key issue though with HHSRS and HMO legislation is that resources are available to the local authority to implement and enforce these.

High cost measures

As will be seen from Table 6.5, many of the measures needed in future will be more expensive than cavity and loft insulation, which have been the key measures to date. In some cases these higher cost measures, such as heat pumps, solar, thermal and solid wall insulation, have not been widely used (at least in the UK). For others, the measures involve fairly new renewable technology. Costs could, at times, be £5,000-£10,000 per dwelling. The importance of these newer measures raises four issues:

- First, programmes and arrangements need to be developed, probably within existing schemes, for installing measures such as small-scale renewables and solid wall insulation in the homes of low-income households where appropriate. A part of the Low Carbon Buildings Programme should be focused on low-income households. In addition, programmes for developing and installing non-mainstream measures (e.g. microgeneration technology) for social housing should be implemented.
- Second, research and development in solid wall insulation is needed so that more cost-effective and customer-friendly methods can be developed.
- Third, it will be important to focus higher cost measures more intensively on those in fuel poverty. This is not straightforward as fuel poverty is caused by a combination of economic (i.e. income) and housing/heating-related inequalities. Local authorities and others with a stake in local communities have a key role in identifying/profiling areas with high incidence of low incomes and energy-inefficient housing.

- Fourth, if significant sums have to be spent on a single owner-occupied dwelling, a contribution should be sought from the householder, for example through an equity release scheme. There are arguments on both sides. On the one hand, the house will often have considerable capital value and in the case of older occupiers their families will often benefit from the measures and any consequent capital appreciation. On the other hand, the value of the homes of those in fuel poverty is, as would be expected, below average (around £130,000 in 2004 compared to around £165,000 for the non fuel poor). More importantly the requirement for the contribution would be a significant disincentive especially for older people.

It is highly likely that equity release will need to make a contribution though. With the introduction of regulation, some of the earlier problems and abuse in the equity release schemes appear to have been resolved. Work still needs to be done on the transaction costs, which can be high and which are a significant deterrent to the release of relatively small sums such as those needed for fuel poverty purposes.

It seems probable that some of the energy companies and/or Eaga will develop equity release schemes in partnership with others to help in fulfilling their programmes and obligations.

There is also likely to be competition from other sectors (e.g. social care) to fund provisions through equity release, so it will be important for schemes to be developed specifically to address housing/fuel poverty/energy-efficiency issues as soon as possible in order to stay ahead of the game.

Locating customers at risk of fuel poverty

The government should, in FPAG's view, take a more holistic approach to identifying, locating and referring customers in fuel poverty who could benefit from the various programmes. Local authorities, NHS or the Department of Work and Pensions will be in touch, one way or another, with most of the targeted households. For this reason, FPAG has proposed to the government that they adopt a proactive, integrated approach using the numerous contacts and the data available to find and help those in fuel poverty. This is very much in line with the government's objective to transform government and provide a more seamless service for customers.

The role of different organisations in tackling fuel poverty

It will be clear from this chapter that the task of combating fuel poverty is a cross-government and cross-organisational one involving a wide range of stakeholders. This is not surprising since, as noted earlier, fuel poverty is influenced by energy efficiency, energy prices and incomes.

The primary responsibility for the fuel poverty target in England is shared by DEFRA (with its energy-efficiency and environmental responsibilities) and the Department for Business, Enterprise and Regulatory Reform (BERR) with its energy portfolio.

The Department for Work and Pensions (DWP), CLG and the Department of Health all have an important role too. Treasury here, as elsewhere, is key – and not just in decisions about expenditure, but in a broader sense (i.e. issues about the role of the market and the extent of intervention).

Ofgem, the energy regulator, potentially has a significant influence on energy prices and, as noted, local authorities (and housing associations) can make significant contributions to reducing fuel poverty.

The six major energy companies have obligations to improve the energy efficiency of their customer base, including special obligations for the homes of low-income (priority group) customers. Their pricing policies are also of great importance and their social programmes can be helpful too. For example, a number of energy companies have set up trusts to help customers with severe difficulties to solve their immediate problems and to find sustainable solutions to some of their financial difficulties.

FPAG is appointed by DEFRA and BERR ministers. They appoint the chairman and nominate the organisations who have representatives on the group. The group has a very broad membership, with representatives from energy companies, NGOs (such as National Energy Action and Help the Aged), consumer organisations like Citizens' Advice and Energywatch, as well as specialist housing bodies such as CIH.

FPAG's role is to advise government on the practical measures needed to meet its fuel poverty targets. It has, for instance, provided estimates of the resources required through the fuel poverty and other programmes to meet these targets.

Targets like the fuel poverty target – that require actions from different departments and agencies – clearly pose significant challenges. The existence of a statutory target, which government is obliged in law to meet, has helped to secure engagement from some departments and agencies, but from by no means all of them. DWP has become much more active on fuel poverty issues, but it has been extremely difficult to engage CLG. Ofgem – a key player – has some useful social projects, but has not focused adequately on fuel poverty, especially in relation to energy price issues.

There are clear lessons here for the Climate Change Bill. If the climate change targets being set out in the bill are to be met, it will be necessary to provide mechanisms to secure the necessary contributions from a range of government departments and agencies. FPAG has tried to raise the visibility of fuel poverty across government and has made policy proposals to each of the relevant departments. Whilst FPAG has made some difference there is still a huge amount to be done.

Conclusions

Houses that require a large amount of energy use are damaging from a climate change point of view and can plunge occupants into fuel poverty. It is therefore of paramount importance to improve the energy efficiency of the existing housing stock as well as to promote the use of household-scale renewables and non-fossil fuels (see existing homes Chapter 10 for an extended discussion). This applies equally to the homes of low- and high-income households. However, special programmes are needed for the homes of low-income households, in order to eradicate fuel poverty, as the household will often not be able to afford the necessary capital measures.

A great deal has been done through the fuel poverty and energy-efficiency programmes to improve the energy efficiency of the homes of low-income customers. However, there is still a very large amount to be done and, in spite of there being a statutory target to eradicate fuel poverty by 2010 amongst vulnerable households, there will be many households which cannot afford to heat their homes adequately. Hence, the 2010 target will not be met, as the government in its latest annual fuel poverty progress report (BERR, 2007) has in effect admitted. This is important in itself, but it also has significant implications for the Climate Change Bill with its statutory targets. The existence of a statutory fuel poverty target has made a difference, has secured more resources for improving the homes of vulnerable people and has resulted in more engagement across government. But it has not made enough difference and it will be important not to expect too much from the climate change statutory targets on their own. A great deal of additional work will be needed to put in place the actual measures required to meet these targets.

References

BERR (2005) *Fuel Poverty 2005 – Detailed Tables*, Department for Business, Enterprise and Regulatory reform. Available at: http://www.berr.gov.uk/files/file42705.pdf

BERR/DEFRA (2007) *The UK Fuel Poverty Strategy: 5th Annual Progress Report*. Available at: http://www.berr.gov.uk/files/file42720.pdf

CLG (2006) *A Decent Home: Definition and Guidance for implementation*. Available at: http://www.communities.gov.uk/documents/housing/pdf/138355

FPAG (2007) Fifth Annual Report. Available at: http://www.berr.gov.uk/files/file38873.pdf

Useful resources

BERR's Design and Demonstration Unit for information on work on gas network extensions and community renewables: http://www.dti.gov.uk/energy/fuel-poverty/ddu/index.html

Climate Change Bill – consultation and supporting documents: http://www.defra.gov.uk/corporate/consult/climatechange-bill/

DEFRA – Domestic energy supplier targets: http://www.defra.gov.uk/environment/climatechange/uk/household/eec/

Eaga Charitable Trust: http://www.eaga.com/charitable/charitable_trust.htm

FPAG: http://www.berr.gov.uk/energy/fuel-poverty/fpag/index.html

Ofgem: http://www.ofgem.gov.uk

Warm Front: http://www.warmfront.co.uk

Warm Zones: http://www.warmzones.co.uk/

CHAPTER 7:
Evolution of policy and practice in low-energy housing

Heather Lovell

Introduction

It was during the late 1990s that climate change began to have a strong influence on the UK housing sector, and low-energy housing policies and practices were reframed as 'low-carbon'. This chapter examines innovation in low-energy housing during this critical period from the late 1990s into the early twenty-first century, focusing on how policy and practice evolved in response to climate change. The pattern of change was for innovation to be concentrated within certain pioneering new-build, low-energy housing developments, such as Hockerton and BedZed. These housing developments also became a focal point of policy debate. The evolution of low-energy housing practices ran ahead of policy, and the UK government struggled to catch up. The overall pattern and process of change fits well with models of innovation based on new ideas and technologies being trialled at a small-scale level, within so-called 'innovation niches' (Schot, Hoogma *et al.*, 1994; Rip and Kemp, 1998), in this case, specific housing developments. The process also corresponds with policy theories about the role of shared discourse and 'storylines' in bringing about change (Hajer, 1995). However, the response of the UK housing sector to climate change during this period offers some challenges to theories about innovation, in particular it raises questions about how effectively learning has taken place from pioneering low-energy housing developments or 'innovation niches'.

The chapter is based on the findings of a three-year research project examining the development of low-energy housing in the UK. Semi-structured interviews were conducted with over 70 experts involved in low-energy housing from a range of housing tenures (social, private and self-build), non-governmental organisations and government. Detailed qualitative case studies of several low-energy housing developments were researched to explore in-depth questions regarding why low-energy housing has been built. The research concentrated on new-build housing – not the existing housing stock – and new-build housing is the focus of analysis here.

The chapter is in three parts. First, theories about technological innovation and policy change are briefly reviewed. Second, two pioneering low-energy housing developments, or innovation niches, are discussed: the BedZed housing development in south London, and the Hockerton Housing Project near Newark in the East Midlands. Third, the pattern and process of change in low-energy housing around the turn of the century is critically examined, concentrating on two issues: low-energy housing practitioners taking action on climate change in advance of policy-makers, and learning from innovation niches.

Ideas about technology innovation and policy change

The literature on technology innovation and policy change is considerable, and it is not intended to review it fully here (see Ham and Hall, 1993, for a review of policy theory; and Jasanoff, Markle et al., 1995, for an overview of technology innovation). Instead, I want to briefly discuss key concepts of technological innovation and policy change that are particularly relevant to the case of low-energy housing. From the discipline of science and technology studies I examine what is known as 'socio-technical transition theory' and the associated model of Strategic Niche Management. From political science I review the concept of policy networks as well as ideas about policy learning. Hitherto policy innovation and technology innovation have tended to be analysed separately (for exceptions see Aibar and Bijker, 1997; Evans, Guy et al., 1999); here I strive to bring together these ideas, because, as the case of low-energy housing demonstrates, the distinction between processes of innovation in policy and in technology is a largely artificial one.

Theories about innovation from science and technology studies

Low-energy housing involves many interconnected innovations – both social and technical – ranging from insulation and solar panels to planning regulations and new householder practices. As such, it is instructive to think of housing as a type of 'socio-technical system', akin to energy, transport or water (see Graham and Marvin, 2001; Osborn and Marvin, 2001; Smith, 2004). The idea of a socio-technical system is central to theories about innovation from science and technology studies: these systems, or regimes, comprise '...[a] rule-set or grammar...embedded in institutions and infrastructures' (Rip and Kemp, 1998: 338). Well-established, durable institutions and infrastructures favour stability, and thus there is a tendency for socio-technical systems to alter mainly through incremental innovations, a characteristic variously described as: 'momentum' (Hughes, 1983; Davies, 1996), 'path dependency' (Phillimore, 2001), 'technological lock-in' (Schot, Hoogma et al., 1994; Unruh, 2002) and entrapment' (Walker, 2000). The main difficulty for those trying to affect socio-technical system change is in overcoming this momentum. It is proposed that radical change therefore tends to start off at a small scale in relatively protected 'innovation niches' (Kemp, Schot et al., 1998; Rip and Kemp, 1998; Smith, 2007). In most cases it is entrepreneurial individuals, businesses and independent researchers who initiate these niches, rather than government, because they are seen as more adept at developing new ideas and have the required flexibility to pursue them (Hughes, 1983).

With the rise of environmental problems such as climate change the focus of technology policy analysis has shifted to consider how governments can catalyse fundamental system-wide change so whole sectors become more environmentally sustainable (Kemp, 1994; Berkhout, 2002; Smith, 2003). In other words, there has been a shift away from a focus on encouraging discrete clean technologies to how to create opportunities for green socio-technical systems. Several models and strategies have been proposed about how governments can play a role in sustainable transitions. Many of these ideas originate from researchers in the Netherlands, where the Dutch government has worked closely with universities to develop new theory and practical tools to manage sustainability transitions. Strategic Niche Management is one such model which provides a framework for how governments can help new radical innovations develop and expand from a base of experiments or niches (Kemp, 1994; Schot, Hoogma *et al.*, 1994; Kemp, Schot *et al.*, 1998; Rip and Kemp, 1998; Smith, 2003; Weber, 2003), defined as '...*the orchestration of the development and introduction of new technologies through setting up a series of experimental settings (niches) in which actors learn about...design, user needs, [and] cultural and political acceptability...*' (Schot, 1992: 261). Strategic Niche Management is a normative, prescriptive approach, and it is perhaps no coincidence that few examples are discussed in the literature, as in practice such well-planned, long-term management is rare. For instance, one case study – the role of the Californian government in promoting electric vehicles – has been examined by a number of authors (see Kemp, 1994; Schot, Hoogma *et al.*, 1994; Schot and Rip, 1996; Rip and Kemp, 1998). Further, it is assumed that governments are able to make strategic decisions about system change; that they have the power and political will to do so (Schot, Hoogma *et al.*, 1994). Smith (2003) rightly questions this assumption, suggesting that because governments tend to be deeply embedded within socio-technical systems they face difficulties in bringing about radical changes, and policies are therefore typically aimed at encouraging incremental innovations.

Theories about innovation from political science

In comparison to theories about technological innovation, there are two main differences in the approach of policy theorists. First, as one might expect, policy theories are principally about government, and there is much less attention to other types of actor and their role in innovation (see for example Marsh and Rhodes, 1992; Sabatier and Jenkins Smith, 1993). As discussed below, in the UK new low-energy housing practices emerged in advance of policy, and the situation therefore does not fit well with this assumption. Second, policy theories tend to embrace the messiness of the process of change and innovation more wholeheartedly than socio-technical theories. For example, the policy theorist John Kingdon likens the policy process to rubbish flowing in and out of a 'garbage can' in that 'streams' of problems, policies and politics co-exist independently, occasionally merging to form a coherent policy programme (Kingdon, 2003). The process of policy change is conceptualised as largely chaotic and non-linear.

But there are also more structured policy theories. There is a long tradition in policy studies of theorising change as driven by networks of actors – termed policy networks – groups of people involved in the policy process from government, corporations, the media and non-governmental organisations (Marsh and Rhodes, 1992; Sabatier, 1999). Policy networks are seen as an important influence on the direction and pace of innovation because it is through these networks that decisions are made on critical issues such as the regulatory environment, grant funding and consumer markets for new innovations (Van De Ven, Polley et al., 1999). There are different ideas about the type of 'glue' that links policy network members. Two popular policy network theories are the 'advocacy coalition framework' – where network members are linked by shared beliefs and values (Sabatier and Jenkins Smith, 1993) – and 'discourse coalitions' – where members are united by shared language (Hajer, 1995). The 'glue' that binds discourse coalitions is the way they talk about an issue and the metaphors or 'storylines' that they use. Storylines are described as '...the essential discursive cement that creates communicative networks among actors with different or at best overlapping perceptions and understandings' (ibid. 1995: 63). It is storylines, therefore, that facilitate understanding between members of discourse coalitions who do not otherwise have much in common.

Although these two types of policy network – advocacy coalitions and discourse coalitions – are commonly presented in opposition to each other – as alternative theories (Hajer, 1995; Sabatier, 1999) – there is emerging evidence that the type of policy network dominating any particular policy sector tends to evolve over time, typically from a single value-based advocacy coalition to multiple discourse coalitions (Lounsbury, Ventresca et al., 2003; Lovell, 2004). This is the pattern observed in UK low-energy housing around the turn of the century, where specific pioneering housing developments were initially developed by people with strong environmental values (akin to an advocacy coalition), and these housing developments were subsequently adopted as 'storylines' for a growing number of people – a discourse coalition – involved in low-energy housing in the late 1990s in response to climate change.

Theories about learning

Where science and technology studies and policy theories about innovation and change are perhaps most closely aligned is through ideas about learning. There is a common assumption that learning takes place through doing (i.e. the process of innovating), and also through seeing innovations in action and directly experiencing them. Case studies (in policy language) or innovation niches (in science and technology studies language) are therefore central to the notion of learning. Learning is seen as both a social and technical process – a point on which policy and technology theorists are agreed. Rose (1991), for example, discusses the role of 'inspirational learning' from seeing new technologies in action. In a similar way Shapin and Schaffer (1985) discuss the role of experimentation in demonstrating the feasibility of new ideas and encouraging others to act; demonstrations reduce the perceived risk of investing in new technologies. A problem shared by both approaches, however, is the scant attention paid to the possible role that case studies or innovation niches might play in *inhibiting* learning. For example, particular projects might be glamorised so that it becomes difficult to criticise them and effectively learn from mistakes. These concerns about the process of learning are returned to later.

Low-energy housing niches

People first began to promote sustainable housing in the early 1970s in the UK (*The Ecologist*, 1972; Bhatti, Brooke *et al.*, 1994; Barton, 1998; Smith, Whitelegg *et al.*, 1998; Chappells and Shove, 2000), concurrent with an increased public awareness of environmental issues, and an upsurge in radical deep green environmentalism (Sandbach, 1980; Weale, 1992; Porter and Brown, 1996; Dryzek, 1997). Examples of environmentally sustainable housing developments from this period include the Centre for Alternative Technology in Wales and the Findhorn Ecovillage in Scotland. There was a shift in the 1990s to sustainable housing developments being built that were typically more focused on energy issues, in response to growing concern about climate change (see Chappells and Shove, 2000). Two examples of this type of energy-oriented housing development, which are the focus of analysis here, are the Hockerton Housing Project near Newark in the East Midlands, and the BedZed development in south London. In these low-energy housing developments dramatic reductions in energy consumption have been achieved, and a number of new technologies and construction methods have been tried out (see Table 7.1 overleaf). These developments include a range of environmental innovations aside from energy or low-carbon ones. For example, at Hockerton there is a reed bed for grey water recycling, and at BedZed there are several innovations aimed at reducing waste, and buying locally (building materials, food, etc.). Nevertheless, with the increasing prominence of the issue of climate change in the 1990s it was the energy features of these developments that were promoted by their developers (note that BedZed is short for 'Beddington Zero Energy Development'), and that received most industry and media attention (Lovell, 2007a).

BedZed is a housing development in south London; the outcome of a joint initiative between the architect Bill Dunster, the Peabody Trust (a housing association), and the environmental consultancy BioRegional Development Group. It consists of 82 homes, nearly half of them sold on the private market, the remainder being social housing. BedZed includes a number of low-energy innovations including an on-site combined heat and power plant, an electric car pool, and sedum grass roofs. Hockerton is an earth-sheltered housing development near Newark in the East Midlands. The five terraced homes have no need for central heating: large conservatories collect heat from the sun, and the walls are very well insulated. Electricity is provided by photovoltaic panels and a wind turbine, and all wastewater is treated on-site in a reed bed. Innovations at Hockerton and BedZed are

Table 7.1: Summary of low-energy features of Hockerton and BedZed housing developments

Characteristic	Hockerton Housing Project	BedZed
Number of dwellings	5	82
Date completed	1998	2000
Type of building material	Concrete, earth sheltered	Masonry
Energy consumption* (kWh/household/year)	c.4,000	c.7,000
Project initiators	Nick Martin (Builder)	Bill Dunster (Architect)
Housing tenure	Self-build (leasehold)	Private and social
Low-energy features	No central heating. Small wind turbine. Photovoltaic (PV) panels; heat pump for hot water; earth-sheltered; passive solar design; 300 mm polystyrene wall insulation; low E window glazing. Low-energy electrical appliances and CFLs.	Passive solar design; no central heating; energy-efficient appliances; on-site combined heat and power station using local wood chips; 300 mm wall insulation; low E window glazing; heat exchangers and passive ventilation.

*Estimated from an average annual energy bill at Millennium Green of £400 (Nash, 2004 personal communication). UK average consumption for home built to the 1995 building regulations is 16,300kWh/household/a (BRECSU, 1996). Sources: BRECSU (2000), BRECSU (2002), Vale (2001), White (2002), Hockerton Housing Project (2003a).

socio-technical, i.e. not purely social or technical, but hybrids, depending on social or personal changes as well as technical ones. Examples include the electric car share scheme at BedZed, and the passive solar design of the dwellings at Hockerton. BedZed and Hockerton have had extensive coverage in specialist and general media in the UK, especially because of their unusual, eye-catching appearance (see Plates 7.1 and 7.2). These two developments have acquired almost a celebrity status with much publicity in mainstream media, not just industry journals.

Both BedZed and Hockerton were initiated by individuals and organisations mostly located outside government. The team responsible for BedZed comprised a mix of organisations including green architects, a housing association and environmental and engineering consultancies. The Hockerton Housing development, a smaller self-build venture, was initiated by an eco-builder – Nick Martin.

Plate 7.1: The BedZed housing development, London (Source: Peabody Trust)

Plate 7.2: Hockerton Housing Project, Hockerton (Source: © Hockerton Housing Project)

Nick Martin was originally contracted to build the Vales' Autonomous House in 1993 in nearby Southwell, and was inspired by that process to become more involved in low-energy housing. The Vales were both trained as green architects and the Autonomous House was the result of their longstanding interest and research on low-energy housing in the UK (Vale and Vale, 2000). Nick Martin owned land on the outskirts of the village of Hockerton, and was encouraged by the Vales (who acted as architects on the project) to build an earth-sheltered housing development, consisting of five terrace houses, completed in 1998.

It would be an over-simplification, however, to say that there was no government influence on BedZed and Hockerton. Local government in particular played a supportive role. For example, in the Newark and Sherwood District Council area, where Hockerton is located, there is an enthusiastic energy manager within the local authority who encouraged low-energy building, making connections between key people and helping to develop progressive local policies. Further, the London Borough of Sutton, where BedZed is located, has a strong environmental policy, and crucially was willing to set a precedent in selling the building land to the BedZed team despite their not bidding the highest price – because of the extra environmental and social benefit it was hoped BedZed would bring.

Analysis of changes in UK low-energy housing policy and practice

I now want to consider in more detail certain features of the way low-energy housing policy and practice evolved during the critical period of the late 1990s, and the implications for policy and theory. Two issues are particularly interesting: first, how innovation was largely confined to what are called 'innovation niches' located outside the policy arena, and second, how effectively learning took place from these niches.

Practice running ahead of policy

A key feature of the evolution of low-energy housing policy and practice in response to climate change in the late 1990s and early 21st century was that changes in practice occurred at a faster pace than changes in policy. In other words, innovation and practical action to mitigate climate change in new-build housing was principally taking place outside government. Like the examples of Hockerton and BedZed, the majority of low-energy housing that was built in the UK during the 1990s was not required by national regulations but instead was built as one-off experimental projects, typically involving an entrepreneurial individual or organisation (Olivier and Willoughby, 1996; Pearson, 1999; see Lowenstein, 2001; Vale, 2001; Synge, 2002). By the early 21st century, however, the UK government was struggling to curb greenhouse gas emissions, and facing increasingly vocal criticism by environmental groups on its low-energy housing policy (or lack of it). It was then that it began to use many ideas from pioneering housing developments such as Hockerton and BedZed as the basis for new policies.

The situation fits closely with ideas from science and technology studies about how early-stage innovation is typically driven by entrepreneurs, businesses and technical enthusiasts at a small scale. The pattern of low-energy housing development also has parallels with policy theories about networks of actors and how they drive change. Early groups of people involved in low-energy housing (from the 1970s through to the 1990s) developed housing based on small, self-sufficient and self-governing communities in line with their environmental values (see for example Wood, 1990; European Eco-village Network, 2003). This initial low-energy housing network in the UK is therefore best characterised as an 'advocacy coalition', united by shared environmental and social values (Shepherd, 2002). It was during the late 1990s that a critical transition took place in low-energy housing in response to climate change. As other more mainstream actors began to be involved – including government – a low-carbon discourse coalition began to dominate. There was an associated shift from a focus on practice (building low-energy homes on a one-off basis) to policy (how to 'mainstream' low-energy housing) (Lovell, 2004). The low-carbon discourse coalition that emerged was a relatively fluid, loose-knit group of actors in comparison to the long-standing sustainable housing advocacy coalition.

So, despite the government not being involved in the initial development of low-energy housing niches, in the late 1990s it grew more interested in them and a number became intimately bound up with UK low-energy housing policy and politics. Low-energy housing developments such as Hockerton and BedZed formed an attractive focus for government for a number of reasons:

- First, because the housing was already there, it was easier for government to make it appear that rapid progress had been made on addressing climate change. This was particularly critical for a government under growing pressure to take climate change action, with evidence emerging of a potential gap between government climate change targets and predicted results from its programmes (RCEP, 2000).
- Second, existing low-energy housing demonstrated that new low-carbon technologies worked, and thus reduced risk for government. The niches grounded the idea of 'low-carbon' development in the housing sector, and thus lent the government (and others) greater credibility.
- Third, because the low-energy housing emerged largely from outside government, it was thus relatively free of associations with past policy programmes, and therefore easier to position and promote as a solution to climate change.

In sum, the low-energy housing niches provided examples of 'ready-made' solutions to climate change that helped the government to communicate its aims, and that it could use to demonstrate that progress was taking place.

It is important, however, not to overstate the role of these particular housing developments in shaping government policy. There were other more general factors influencing the development of low-energy housing policy at the same time, ranging from concerns about an undersupply of housing (DTLR, 2000; Barker, 2003), to growing public support for action on climate change (Newell, 2000; Dryzek, Downes *et al.*, 2003; Smith, 2004), and forthcoming European regulations on energy use in buildings (EC, 2002). I want to argue a more modest case for how these particular schemes played a role in encouraging government to take action, and also focused efforts in certain ways.

A key approach of government in trying to associate itself with high-profile low-energy developments was to present them as case studies in government reports. Table 7.2 gives examples of UK policy documents citing BedZed. The policy documents neglect to mention that BedZed emerged without any significant government support. The government nevertheless attempted to claim ownership of BedZed through frequent references to it, and through site visits, including the launch of new policies there (BRECSU, 2002; DTI, 2002; DTI, ODPM *et al.*, 2003; The Stationery Office, 2003; Watts, 2003; The Housing Corporation, 2004). For example, Patricia Hewitt, the then Secretary for Trade and Industry, used BedZed to announce a new government solar power initiative (DTI, 2002). Similarly, the Liberal Democrat party leader visited because (as I was told by someone involved) he *'...was making an environment announcement later that day and wanted a photo to go with any publicity'.*

Table 7.2: Examples of UK policy documents citing BedZed

Policy document	Reference to BedZed
Speech by Energy Minister Brian Wilson, February 2002 (DTI, 2003b).	*'Demonstrations such as the developments... at BedZed...prove that the technologies are available to deliver practical systems.'*
Royal Commission on Environmental Pollution 22nd Report: *Energy – Our Changing Climate.*	Has a case study box devoted to BedZed and describes it as: *'...the most ambitious low energy housing development in the UK to date...'* (RCEP, 2000: 105).
Government Energy Efficiency Best Practice Programme – General Information Report no. 89.	*'...BedZed represent[s] state-of-the-art for sustainable housing in the UK'* (BRECSU, 2002: 3).
UK 2003: the Official Yearbook of the United Kingdom of Great Britain and Northern Ireland (The Stationery Office, 2003).	Double page picture spread (pp.298-299).
Environment Agency report – *Our Urban Future,* September 2002 (www.environment-agency.gov.uk).	BedZed is cited as an example of a solution to climate change.
The Housing Corporation (2004).	It is used as a model case study for Registered Social Landlords *'...to show how sustainable development can be achieved'.*

In this way BedZed became a key part of what the government said about low-energy housing, acting effectively as a 'storyline', as did Hockerton (Lovell, 2007b). The low-energy housing niches were a focal point for policy makers, uniting an otherwise disparate expanding group of actors trying to address climate change within the housing sector. The notion of a storyline is intimately connected with discourse coalition theory (Hajer, 1995; 2001), although the idea of new technologies or innovations acting as storylines is a relatively new one (see Lovell, 2007b). The low-energy housing storylines have been effective in generating industry and government support for low-energy housing innovations, and facilitating market demand. For example, the authors of a government-commissioned report about BedZed stress how the development represents:

> '...a powerful argument for the feasibility of a zero-carbon target for all new build' (BRECSU, 2002: 11).

And in a government-sponsored case study of the successful energy policies and programmes within the Newark and Sherwood District Council, where Hockerton is located, one of the lessons learnt is that:

> '...exemplar projects bring to life the reality far greater than shelves of strategies' (Energy Saving Trust, 2004: 10).

As well as drawing on low-energy housing niches in policy discourse and government reports, the government more substantially translated ideas and innovations from the developments into policy. For instance, findings from government reports about Hockerton and BedZed (BRECSU, 1996; 2000; 2002) influenced changes to the building regulations (ODPM, 2000; 2003). In addition, new policies have been forthcoming at a local government level, based on the experience of BedZed and Hockerton. Sutton Borough Council, where BedZed is located, set an important precedent by awarding the development contract to the BedZed team, despite not being the highest bidder, because of its environmental advantages. Experience with BedZed also influenced the Unitary Development Plan in neighbouring Merton Borough Council, which required new developments over a certain size to source ten per cent of their energy from renewable resources (Forum for the Future, 2004). This so-called 'Merton Rule' was subsequently adopted by local authorities across the UK.

As well as drawing lessons from existing niches, the government drew up policies aimed at promoting new innovation niches, typically through competition for grants for low-energy projects (see Table 7.3). For example, the government's Low Carbon Buildings Programme aimed to '...demonstrate on a wider scale emerging micro generation technologies...' and '...to raise awareness by linking demonstration projects to a wider programme of activities...' (Low Carbon Buildings Programme, 2006). Eighty million pounds was made available, with funding streams directed at householders and community groups, as well as public buildings and businesses.

Table 7.3: UK government grants and programmes promoting the development of new low-energy housing developments

Name of grant/ programme	Date	Details
Community Renewables Initiative	2002- present day	Local communities bid for funding for renewable energy projects. Funding from DTI. (see The Countryside Agency, 2004)
Clear Skies	2003-2005	Capital grants for household and community renewable projects. (see BRE, 2003)
Solar PV Programme	2002-2005	£20 million budget was available. Stream One of funding designed for home owners and small businesses, Stream Two for community bids (and large public buildings, businesses etc.). (see DTI, 2003a)
Community Energy Programme (CHP)	2002-2007	For combined heat and power and district heating technologies only. (see Energy Saving Trust and The Carbon Trust, 2001)
Scottish Community and Householder Renewables Initiative (SCHRI)	2002- present day	Offers funding to householders and community groups for installing renewable energy technologies in Scotland. (see Energy Saving Trust, 2006)
DTI's Low Carbon Buildings Programme (LCBP)	2006	Designed to replace Clear Skies and the PV demonstration programme. Minimum energy-efficiency measures must be undertaken in order to qualify for a renewable energy grant – aims at a more holistic low-energy approach than previous government programmes. (see Low Carbon Buildings Programme, 2006)
Millennium Communities Programme	1997- present day	Seven communities are being developed as examples of housing best practice – including environmental sustainability, e.g. the Greenwich Millennium Village. (see English Partnerships, 2003)

The UK government's focus on niches in part results from its inability to create wider change (see Lovell, 2007b for further discussion). Unruh's (2002) research on climate change policy leads him to conclude that niches might appeal to governments more for political than technical reasons, as he explains:

'For policy makers constrained by [technological] lock-in, but still seeking to provide incentives for carbon saving alternatives, niches become an attractive policy target' (Unruh, 2002: 322).

The idea is that change can be achieved more easily within niches without threatening existing interests within well-established systems such as energy and housing. Unruh's critique of government also offers some explanation as to why the government's response to climate change in the housing sector has tended to focus strongly on new-build housing as opposed to the existing housing stock (for example through the government's zero-carbon target for all new-build housing by 2016).

Learning from low-energy housing niches

As we have said, policy and technology theorists share some common ground, in that both see learning as often taking place from best practice case studies or innovation niches (Sanderson, 2002; Szejnwald Brown, Vergragt *et al.*, 2003; Bulkeley, 2006). This type of learning process is certainly evident in the case of UK low-energy housing. Several instances of learning from pioneering low-energy housing developments have been mentioned, for example the development of new government policies based on experience at BedZed and Hockerton. The housing developments have also promoted learning and further innovation in a more indirect way: the fact that they are there helps to convince others – government, business and householders – of the commercial and technical feasibility of low-carbon housing. As the project manager of the Hockerton housing project puts it:

'The most effective tool is the place and the fact that we are living in it' (Interview, Hockerton Project Manager, March 2003).

And a housing policy officer at a non-governmental organisation involved in sustainable housing similarly describes how they have used BedZed to influence key decision-makers in government:

'I think [exemplar projects] are invaluable for showing people what might be done. It is really great when we want to talk to people about sustainable housing – important people – we take them to BedZed...and to actually see it in action I think is very inspiring, rather than just talking about what it might look like' (Interview, Sustainable Homes co-ordinator at a national environmental NGO, May 2003).

The interviewee thus hints at how, with growing discussion about low-carbon housing, it has become increasingly important not just to participate in the discourse, but to have actual evidence of low-carbon *practice* in order to promote new ideas and gain support. These findings echo the ideas of Shapin and Schaffer (1985) about the role of experimentation being primarily social, as a means of demonstrating new ideas to others – 'generating facts' – and gaining acceptance for them.

However, there are some problems with low-energy housing developments being seen in this way, i.e. as storylines or metaphors rather than practical learning experiments. First, the storylines have prioritised low-energy technologies over other types of sustainable innovation within these developments, such as water or waste technologies, which have had far less publicity and attention. Second, the way these housing developments have been uncritically portrayed by government and in the media has inhibited learning from them. As already mentioned, BedZed and Hockerton have both had extensive coverage in specialist and general media. This coverage has almost all been positive: 99 per cent of the articles written about BedZed were complimentary (BioRegional Communications Officer, 2004), as were all but two of the articles or programmes about Hockerton. This positive publicity may in part be a reflection of the short time between the completion of the developments and the glut of media and government attention; it can take years for problems with housing developments to become evident. Indeed, more recent media treatment has been critical, identifying a number of problems with BedZed in particular (see for example, Slavin, 2006; Steffen, 2006). Nevertheless, it is the case that good publicity is central to maintaining the positive policy storylines about these developments, and there is some evidence to suggest that initial project problems with BedZed and Hockerton were stifled for this reason. For example, the combined heat and power plant at BedZed has had ongoing problems, and the project is said to have overrun its budget by £5 million (Clark and Smit, 2004; Lovell, 2005). Yet there is no hint of these problems in a government-commissioned review of BedZed, in which the development is uncritically applauded as representing:

'...state-of-the-art for sustainable housing in the UK' (BRECSU, 2002: 3).

Similarly at Hockerton, some of the houses have experienced low winter temperatures, falling to below 18 degrees centigrade in winter months, and extra heating in the form of oil radiators and wood stoves has been used. A lack of honesty about problems with existing low-energy housing technologies can lead to insufficient learning, and a risk of the technologies being used in a more widespread manner without the problems being corrected. Rose (1991) has suggested that honest accounts of policies or projects are not necessary for learning, because the main impact of demonstration projects is to generate 'inspirational learning', thereby motivating others to adopt similar approaches. However, Rose's focus is more on learning about social issues than technical ones: there are perhaps greater limitations to inspirational learning if significant technical problems are replicated. Indeed there is little sense from government policy about low-energy housing that the niches are primarily seen as technical experiments (Lovell, 2004). So, although it is acknowledged in the literature on strategic niche management that niches are for social as well as technical learning, in these instances any technical learning has been eclipsed by the political promotion of the niches as examples for others to follow. For example, in a government report examining four low-energy housing niches, it is explained in the introduction how:

*'These case studies demonstrate the successful integration of renewable energy into new housing projects...[they] should offer **reassurance** and **inspiration** to building designers, consultants and anyone involved in the specification and design of dwellings'* (BRECSU, 2003: 1, emphasis added).

Thus, although the report is ostensibly technical – it was written and produced by the Building Research Establishment – it lacks detailed analysis of the performance of the housing developments, including information about any technical problems. One or two contentious issues are raised, such as contractors lacking confidence and experience in installing renewable energy technologies (BRECSU, 2003), but nothing substantial. The overall tone equates to that of a marketing brochure: the existing housing developments are being promoted as best practice case studies to building industry professionals to encourage them to take similar action. There is little sense that an independent, robust technical analysis has been conducted.

Summary and conclusions

In summary, this analysis of changes in UK low-energy housing policy and practice during the late 1990s and early 21st century has highlighted a number of issues about the process of innovation and change, particularly in the role of government and non-state organisations. First, entrepreneurial individuals and organisations outside government can play an important role in driving policy and technology change. In the case of low-energy housing, non-state organisations (green architects, eco-builders, housing associations, etc.) were helpful to government in a critical period when climate change rose up the policy and political agenda and the government was struggling to come up with a suitable policy response. It was useful for the government to have examples of best practice to draw upon and promote, such as BedZed and Hockerton.

Second, and related, in this period there was a shift in the type of actor involved in low-energy housing in the UK: low energy housing innovations moved from the fringe of the housing sector (pioneered by individual enthusiasts and eco-builders) to mainstream practice, adopted by private sector housebuilders and incorporated into government policy. There was a corresponding change in the type of policy network in operation: from an initial advocacy coalition – with members closely united by shared environmental values – to a discourse coalition – a looser network of people linked through shared language about climate change.

Third, it is evident that the process of policy innovation and learning is messy and unpredictable. The main implications for policy-makers relate to this last point. There is danger in trying to smooth and gloss over problems arising from developing and building new types of low-energy housing. There is a need to have thorough ongoing analysis of pioneering projects to ensure that learning does take place, both technical and social, and that problems are not hidden but openly discussed.

Further, there is more general danger of policy-makers focusing on exceptional pioneering developments, giving the impression that they are easily replicated. As the cases of BedZed and Hockerton demonstrate, there are many unusual features of these developments not least in terms of the people involved, the planning process, and the housebuilding methods and technologies used. Policy attention needs to be directed not just at encouraging more of these innovative developments, but also at addressing why most new housing developments in the UK are not like them.

References

Aibar, E. and Bijker, W. E. (1997) 'Constructing a City: the Cerda Plan for the extension of Barcelona', *Science, Technology and Human Values* 22(1): 3-30

Barker, K. (2003) *Review of Housing Supply – Securing our Future Housing Needs. Interim Report – Analysis*, London: HMSO

Barton, H. (1998) 'Eco-neighbourhoods: a review of projects', *Local Environment* 3(2): 159-177

Berkhout, F. (2002) 'Technological regimes, path dependency and the environment', *Global Environmental Change* 12(1): 1-4

Bhatti, M., Brooke, J. *et al.* (1994) 'Housing and the New Environmental Agenda: an Introduction' in *Housing and the environment: A new agenda*. M. Bhatti, J. Brooke and M. Gibson, Coventry: Chartered Institute of Housing: 1-13

BioRegional Communications Officer (2004) Email correspondence: Jennie Organ, Communications Officer – BioRegional Development Group, H. Lovell, London

BRE (2003) *Clear Skies – Renewable Energy Grants,* Building Research Establishment (BRE). Retrieved 21st May 2003 from: http://www.clear-skies.org

BRECSU (1996) *General Information Report 53: Building a sustainable future – homes for an autonomous community*, Watford, UK: Building Research Establishment (BRE)

BRECSU (2000) *New Practice Profile 119: The Hockerton Housing Project – design lessons for developers and clients*, Watford, UK: Building Research Establishment (BRE)

BRECSU (2002) *General Information Report 89: BedZed – Beddington Zero Energy Development, Sutton*, Watford, UK: Building Research Establishment (BRE)

BRECSU (2003) *Renewable energy in housing – case studies*, Watford, UK: Building Research Establishment (BRE)

Bulkeley, H. (2006) 'Urban sustainability: learning from best practice?', *Environment and Planning A* 38: 1029-1044

Centre for Alternative Technology (1995) *Crazy Idealists? The CAT story*, Machynlleth: The Centre for Alternative Technology

Chappells, H. and Shove, E. (2000) 'Sustainable Homes and Integration' in *Domestic Consumption Utility Services and the Environment (DOMUS)*, H. Chappells, M. Klintman, A.-L. Linden *et al.*, Wageningen: Universities of Lancaster, Wageningen and Lund: 105-128

Clark, P. and Smit, J. (2004) 'Dream Over', *Building:* 1-3

Communities and Local Government (2006) *Building a Greener Future: Towards Zero Carbon Development – Consultation*, London: HMSO

Countryside Agency (2004) *The Community Renewables Initiative*, The Countryside Agency. Retrieved 1st January 2005 from: http://www.countryside.gov.uk/NewEnterprise/Economies/CRI.asp

Davies, A. (1996) 'Innovation in Large Technical Systems: The Case of Telecommunications', *Industrial and Corporate Change* 5(4): 1143-1180

Dryzek, J. (1997) *The politics of the Earth: environmental discourses*, Oxford: Oxford University Press

Dryzek, J. S., Downes, D. *et al.* (2003) *Green States and Social Movements: Environmentalism in the United States, United Kingdom, Germany and Norway*, Oxford: Oxford University Press

DTI (2002) 'First stage of major PV Demonstration Programme launched', *New Review* 52: 1

DTI (2003) *DTI Major Photovoltaics (PV) Demonstration Programme*, Department for Trade and Industry. Retrieved 26th March 2003 from: http://www.est.co.uk/solar

DTI, ODPM, *et al.* (2003) *Better Buildings Summit Issues Paper*, London: HMSO

DTLR (2000) *Quality and Choice – A Decent Home for All*, White Paper, London: The Stationery Office

EC (2002) *Directive 2002/91/EC of the European Parliament and of the Council on the energy performance of buildings.* Available from: http://europa.eu.int/eur-lex/pri/en/oj/dat/2003/l_001/l_00120030104en00650071.pdf

Energy Saving Trust (2004) *Newark and Sherwood District Council case study: a guide for local authorities*, London: Energy Saving Trust for the Energy Efficiency Best Practice in Housing Programme

Energy Saving Trust (2007) *Housing and Buildings Funding Database*. Retrieved 18th March 2007 from: http://www.energysavingtrust.org.uk/housingbuildings/funding/database/

Energy Saving Trust, The Carbon Trust (2001) *Community Energy Programme*, DEFRA. Retrieved 11th November 2003 from: http://www.communityenergy.org.uk

English Partnerships (2003) *Millennium Communities Programme*, English Partnerships. Retrieved 21st May 2003 from: http://www.englishpartnerships.gov.uk

European Eco-village Network (2003) *What are ecovillages?*. Retrieved 23rd April 2003 from: http://europe.ecovillage.org/uk/network/index.htm

Evans, R., Guy, S. *et al.* (1999) 'Making a Difference: Sociology of Scientific Knowledge and Urban Energy Policies', *Science, Technology and Human Values* 24(1): 105-131

Findhorn Ecovillage (2003) *Findhorn – an Introduction*. Retrieved 23rd April 2003 from: http://www.ecovillagefindhorn.com/

Forum for the Future (2004) *Merton Borough Council – a radical planning decision to promote renewable energy.* Retrieved 24th August 2004 from: http://www.regionalfutures.org.uk/newsdigest/mertonboroughcouncil_page2181.aspx

Graham, S. and Marvin, S. (2001) *Splintering urbanism: networked infrastructures, technological mobilities and the urban condition*, London: Routledge

Hajer, M. A. (1995) *The politics of environmental discourse: ecological modernisation and the policy process*, Oxford: Clarendon Press

Hajer, M. A. (2001) *A Frame in the Fields: Policy making and the reinvention of politics*, European Consortium for Political Research Conference (Workshop 9 – Policy Discourse and Institutional Reform), Grenoble, France: http://www.essex.ac.uk/ecpr/jointsessions/grenoble/papers/ws9.htm

Ham, C. and Hall, M. (1993) *The Policy Process in the Modern Capitalist State*, Hemel Hempstead: Harvester Wheatsheaf

Hockerton Housing Project (2003a) *Hockerton Housing Development – Technical Tour*, 20th March 2003

Hughes, T. P. (1983) *Networks of Power: Electrification in Western Society 1880-1930*, Maryland: The John Hopkins University Press

Jasanoff, S., Markle, G. E. *et al.*, (Eds.) (1995) *Handbook of Science and Technology Studies*, London: Sage

Kemp, R. (1994) 'Technology and the Transition to Environmental Sustainability', *Futures* 26(10): 1023-1046

Kemp, R., Schot, J. W. *et al.* (1998) 'Regime shifts to sustainability through processes of niche formation: the approach of Strategic Niche Management', *Technology Analysis and Strategic Management* 10(2): 175-195

Kingdon, J. W. (2003) *Agendas, alternatives and public policies*, New York: Harper Collins College Publishers

Lounsbury, M., Ventresca, M. J. *et al.* (2003) 'Social movements, field frames and industry emergence: a cultural-political perspective on US recycling', *Socio-Economic Review* 1: 71-104

Lovell, H. (2004) 'Framing sustainable housing as a solution to climate change', *Journal of Environmental Policy and Planning* 6(1): 35-56

Lovell, H. (2005) *The governance of emerging socio-technical systems: the case of low energy housing in the UK*, Cambridge: Cambridge University, Department of Geography

Lovell, H. (2007a) 'Exploring the role of materials in policy change: innovation in low energy housing in the UK', *Environment and Planning A* 39(10): 2500-2517

Lovell, H. (2007b) 'The governance of innovation in socio-technical systems: the difficulties of strategic niche management in practice', *Science and Public Policy* 34: 35-44

Low Carbon Buildings Programme (2006) *DTI Low Carbon Buildings Programme – Frequently asked questions.* Retrieved 22nd April 2006 from: http://www.est.org.uk/housingbuildings/funding/lowcarbonbuildings/faq/

Lowenstein, O. (2001) 'From BedZed to eternity', *Building for a Future* 11: 16-21

Lowenstein, O. (2001) 'Zero energy homes come to the United Kingdom', *Building for a Future* 11: 8-14

Marsh, D. and Rhodes, R. A. W., (Eds.) (1992) *Policy Networks in British Government*, Oxford: Clarendon Press

Newell, P. (2000) *Climate for Change: non-state actors and the global politics of the greenhouse*, Cambridge: Cambridge University Press

ODPM (2000) *Current thinking on possible future amendments of energy efficiency provisions.* Retrieved 23rd February 2005 from: http://www.odpm.gov.uk/stellent/groups/odpm_buildreg/documents/page/odpm_breg_600333.hcsp

ODPM (2003) *Industry Advisory Group (Working Party 4) Low and Zero Carbon Energy Systems Meeting Minutes – 16th October 2003.* Retrieved 25th April 2004 from: http://www.odpm.gov.uk

Olivier, D. and Willoughby, J. (1996) *General Information Report no.38: Review of ultra-low-energy homes – a series of United Kingdom and overseas profiles*, Watford: Building Research Establishment

Osborn, S. and Marvin, S. (2001) 'Restabilizing a Heterogeneous Network: the Yorkshire Drought 1995-96' in *Infrastructure in transition: urban networks, buildings, plans*, S. Guy, S. Marvin and T. Moss, London: Earthscan: 68-77

Pearson, A. (1999) 'Green fingers', *Building*: 30-31

Phillimore, J. (2001) 'Schumpeter, Schumacher and the Greening of Technology', *Technology Analysis and Strategic Management* 13(1): 23-37

Porter, G. and Brown, J. W. (1996) *Global Environmental Politics*, Boulder: Westview Press Inc

RCEP (2000) *Energy – The Changing Climate, 22nd Report*, London: HMSO

Rip, A. and Kemp, R. (1998) 'Technological Change' in *Human Choices and Climate Change Volume 2 – Resources and Technology*, S. Rayner and E. Malone, Columbus, Ohio: Battelle Press: 327-399

Rose, R. (1991) 'What is lesson drawing', *Journal of Public Policy* 11(3): 3-30

Sabatier, P. A. (1999) *Theories of the Policy Process: theoretical lenses on public policy*, Boulder: Westview Press

Sabatier, P. A. and Jenkins Smith, H. C., (Eds.) (1993) *Policy change and learning: an advocacy coalition approach*, Boulder: Westview Press

Sandbach, F. (1980) *Environment, Ideology and Policy*, Oxford: Basil Blackwell

Sanderson, I. (2002) 'Evaluation, policy-learning and evidence-based policy making', *Public Administration* 80(1): 1-22

Schot, J. W. (1992) 'Constructive Technology Assessment and Technology Dynamics: the case of clean technology', *Science, Technology and Human Values* 17(1): 36-56

Schot, J. W., Hoogma, R., *et al.* (1994) 'Strategies for Shifting Technological Systems: the case of the automobile system', *Futures* 26(10): 1060-1076

Schot, J. W. and Rip, A. (1996) 'The Past and Future of Constructive Technology Assessment', *Technological Forecasting and Social Change* 54(2/3): 251-268

Shapin, S. and Schaffer, S. (1985) *Leviathan and the Air Pump: Hobbes, Boyle and the Experimental Life*, Princeton: Princeton University Press

Shepherd, N. (2002) 'Anarcho-environmentalists: ascetics of late modernity', *Journal of Contemporary Ethnography* 31(2): 135-157

Slavin, T. (2006) 'Living in a Dream', *The Guardian*, London: The Guardian: 6

Smith, A. (2003) 'Transforming Technological Regimes for Sustainable Development: a role for Alternative Technology niches?', *Science and Public Policy* 30(2): 127-135

Smith, A. (2004) *Governance lessons from green niches: the case of eco-housing*, ESRC Sustainable Technologies Programme workshop on Governance, Technology and Sustainability, Milton Keynes: Open University

Smith, A. (2004) 'Policy transfer in the development of UK climate policy', *Policy & Politics* 32(1): 79-93(15)

Smith, A. (2007) 'Translating sustainabilities between green niches and socio-technical regimes', *Technology Analysis & Strategic Management*

Smith, M., Whitelegg, J. *et al.* (1998) *Greening the Built Environment*, London: Earthscan

Steffen, A. (2006) *Trouble at BedZed*. Retrieved 19th February 2008 from: http://www.worldchanging.com/archives//004454.html

Synge, D. (2002) 'Welcome to the green living machine', *The Financial Times*, London: The Financial Times: 6

Szejnwald Brown, H., Vergragt, P. *et al.* (2003) 'Learning for sustainability transition through bounded socio-technical experiments in personal mobility', *Technology Analysis & Strategic Management* 15(3): 291-316

The Ecologist (1972) 'Blueprint for Survival', *The Ecologist*, London: The Ecologist

The Housing Corporation (2004) *BedZed case study for Housing Associations*. Retrieved 22nd October 2004 from: http://www.housingcorp.gov.uk/resources/sustain.htm#tools

The Stationery Office (2003) *The official yearbook of the United Kingdom of Great Britain and Northern Ireland*, London: HMSO

Toke, D. (2000) 'Policy network creation: the case of energy efficiency', *Public Administration* 78(4): 835-854

Unruh, G. C. (2002) 'Escaping carbon lock-in', *Energy Policy* 30(4): 317-326

Vale, B. and Vale, R. (2000) *The New Autonomous House*, London: Thames & Hudson

Vale, R. (2001) *Hockerton Housing Project – Case Study 7.7b*, Canberra: Australian

Greenhouse Office, Institute for Sustainable Futures: 1-6

Van De Ven, A. H., Polley, D. E. *et al.* (1999) *The Innovation Journey*, New York & Oxford: Oxford University Press

Walker, W. (2000) 'Entrapment in large technology systems: institutional commitment and power relations', *Research Policy* 29(7): 833-846

Watts, C. (2003) 'Housing for the Future', *The Source Public Management Journal*, 13th July 2002: 1-2

Weale, A. (1992) *The new politics of pollution*, Manchester and New York: Manchester University Press

Weber, K. M. (2003) 'Transforming Large Socio-technical Systems towards Sustainability: on the Role of Users and Future Visions for the Uptake of City Logistics and Combined Heat and Power Generation', *Innovation: the European Journal of Social Science Research* 16(2): 155-175

White, N. (2002) *Sustainable Housing Schemes in the United Kingdom*, Hockerton: Hockerton Housing Project

Wood, A. (1990) *History and overview of communal living.* Retrieved 22nd April 2003 from: http://www.diggersanddreamers.org.uk/Articles/199001.htm

CHAPTER 8:
Building new houses, are we moving towards zero-carbon?

Melissa Taylor and Tom Woolley

'The strategic housing decisions we take collectively over the next few years are critical to the life chances of the next generation' (CLG, 2007a).

Introduction

Domestic activities account for more than 27 per cent of UK carbon emissions (WWF, 2008). Although the vast majority of our housing stock was built more than ten years ago, new build is crucial to reducing those emissions. It is thought that by 2050 one-third of the homes in the UK will have been built since 2007 (CLG, 2007b). More recent estimates put this figure even higher (Boardman, 2007). The UK government has embarked on an ambitious strategy to achieve high standards of energy efficiency in new housing, also requiring the inclusion of renewable energy generation. These plans are encapsulated in the 'Code for Sustainable Homes', however, the standards have not been enthusiastically embraced by all parts of the new-build sector. The private house building industry takes the view that government is rushing ahead too fast, advocating technologies that may not work and may not achieve the zero-carbon objectives (see Box 8.1). The social housing sector, mostly housing associations, voice their concern over the cost implications in adopting innovative solutions when the supply chain is not prepared. Activists from the green movement argue that the new standards are not stringent enough but also fear that many of the new technologies will fail, or lead to problems, or are too costly. There seems to be little consensus as to the best strategies to adopt though, and as the new standards start to bite, inadequate solutions may fall by the wayside.

Box 8.1: Risks threatening delivery of the 2016 target

- **A shorter timescale** would jeopardise the success of work to develop skills and effective and reliable products, with consequent risks for consumer confidence.
- **Unrealistic 'free-for-all' on local targets** would fragment efforts to achieve economies of scale and prevent a concerted focus from the supply chain in developing the most promising new products efficiently.
- Lack of a clear national framework would **complicate efforts to meet skills requirements** and the management burdens for local authorities, developers and others.

→

- **Housing supply could be adversely affected** – either because of supply chain problems or because additional costs would choke off development viability.
- In either case, such consequences would necessarily **worsen the enormous affordability pressures** faced by communities across the country.
- If fewer new homes are built, this is likely in itself to have **adverse environmental consequences**. There will be greater reliance on the much less carbon-efficient existing housing stock and people who cannot live where they need to may well need to make longer car journeys to their place of work.

Source: House Builders Federation: *Zero Carbon Homes – Delivering the 2016 target*, (2007).

National Housing Federation chief executive David Orr has warned that the government is in danger of missing its target to ensure all new homes are zero-carbon by 2016 (NHF, 2007). Speaking at the Federation's annual conference in 2007, Mr Orr revealed that while 92 per cent of housing association new homes are already meeting minimum sustainability standards, only two per cent of new homes built by private developers do so (ibid.).

Whilst even the private sector cannot publicly say they are against reducing carbon emissions, there is serious disquiet as to whether the new standards can be met. Firstly though, we need to discuss what these targets consist of.

Government targets

Zero-carbon new homes by 2016

In order to reduce UK domestic CO_2 emissions, the government has set an ambitious target for new homes to be 'zero-carbon' by 2016. The current government definition of zero-carbon is as follows:

- that a home produces zero net carbon dioxide emissions over the course of a year;
- that this takes into account all the energy used in the house – for lighting, heating, cooling, cooking, running the TV, and so on; and
- that low and zero-carbon solutions can be developed across a site, not necessarily for each individual home (Smith, 2007).

This means that new homes will need to reduce the energy that is used for activities such as space heating, water heating, lighting, cooking and running electrical appliances. Once that energy is reduced to a minimum, homes will need to supply the remaining energy required from a renewable or zero-carbon energy source, so that over a year there are no net carbon dioxide emissions as a result of the activities within the home. Government has also introduced stamp duty land tax exemption of new zero-carbon homes costing less than £500,000. From October 2007, in order to qualify, a zero-carbon home must be certified and calculated in accordance with the approved method, set out in the Stamp Duty Land Tax (Zero Carbon Homes Relief) Regulations 2007 (OPSI, 2007).

Carbon emissions associated with production and transport of building materials are not considered in this definition, nor is the energy used during the construction process. With the construction of each new home thought to be responsible for 20-40 tonnes of CO_2 (Brinkley, 2008), there is growing concern that the present definition of a zero-carbon home does not reflect the 'true' carbon footprint.

It may be argued that the definition of a zero-carbon home should also address carbon emissions associated with the lifestyle of the occupant. The location of the home will greatly influence the mode of transport, how and where food is bought, travel to work and school, and interactions with their local community. All of these activities have associated CO_2 emissions.

Three million new homes by 2020

As well as targeting carbon emissions, the government has also set ambitious targets for increasing housing supply, stating that three million new homes are needed by 2020, with building increasing to a rate of 240,000 per year by 2016 (CLG, 2007a).

The stated aim is for these three million new homes to be of good quality, affordable and sustainable. A number of initiatives have been put in place in order to help make these targets possible.

Sustainable Communities Plan: In 2003, four growth areas were designated as part of the Sustainable Communities Plan. The aim was to tackle supply issues and to create communities that were environmentally, socially and economically sustainable (Shaw, 2007).

Millennium Communities Programme: Nine thousand new homes built to EcoHomes Excellent standards, together with supporting commercial, school, community and health buildings, are expected in the development of seven newly created 'exemplar' sustainable communities (EP, 2007). Over 1,000 homes have already been completed at Greenwich Millennium Village, New Islington in East Manchester, and Allerton Bywater near Leeds.

Eco-towns: In May 2007, the government announced plans to build a number of 'eco-towns'. Each will provide at least 5,000-20,000 new homes, with supporting facilities including a secondary school, shopping, business space and leisure facilities. The developments as a whole are expected to achieve zero-carbon and to be models of sustainable living (CLG, 2007c).

Design for Manufacture Competition – the 60k house

In September 2004 the government launched a competition to build homes for £60,000 (CLG and EP, 2006). Included in the aims of the project were (i) minimum of EcoHomes Very Good rating, (ii) reduction in construction waste and (iii) a drive in culture change within the construction industry, increasing efficiencies and quality. While some housing developers claimed to have met the 60K target, this was widely criticised in the press and the 60K home project has quietly disappeared from view. The programme was even attacked in the House of Commons, though some of the technology referred to has re-emerged in other guises:

> 'The energy efficiency credentials and ability of the winning entries of Government's £60K House competition to meet current building regulations has been questioned by a backbench Labour MP. Alan Simpson, MP for Nottingham South, has strongly criticised the Government's approval of lightweight housing systems for the £60K house. Their construction would provide homes that are too hot in the summer and too cold in the winter. Such homes would require increased air conditioning and heating with consequent higher energy demands and carbon emissions. Simpson described the Government's decision as "a classic example of stealing cheaply from today what we will have to pay dearly for tomorrow". His concerns are echoed by the Royal Institute of Chartered Surveyors who have raised the issue with the environmental minister Elliott Morley' (Concrete Centre, 2006).

Carbon Challenge

The most recent government initiative for housebuilding is specifically aimed at the zero-carbon homes target. The Carbon Challenge, announced in February 2007 aims to '...develop the skills and technologies in the house building industry that are necessary to deliver new zero carbon homes at Level 6 of the Code for Sustainable Homes' (CLG and EP, 2007). The first site to be developed under the scheme is at Hallam Hall, an old hospital site near Bristol, to be followed by a second site in Peterborough city centre.

Construction of zero-carbon houses

A wide range of new technologies are appearing in an effort to meet low and zero-carbon standards, not just for creating renewable energy, but also in terms of constructing new houses. Many of these techniques are hybrids of tried and tested solutions but many are highly innovative. In the main, the new forms of construction are more expensive than so-called traditional building, but the incorporation of much higher specification windows and doors, ventilation systems and substantially more insulation will inevitably increase costs. In an effort to increase efficiency and keep costs within limits there has been encouragement for more prefabrication, off-site construction and other new techniques. These range from so-called 'high-tech' materials using steel and synthetic components, to timber frame and the use of low impact renewable crop-based materials.

Many of these innovations have been encouraged by government, and some are show-cased at the Building Research Establishment's (BRE) Innovation Park and at an annual 'Off-site' conference and exhibition.

These showcase buildings have highlighted major challenges facing housebuilders aiming to achieve zero-carbon performance. Only six per cent of home owners surveyed by the National House Building Council Foundation (NHBC) felt that the £35,000 extra cost to achieve a zero-carbon home is reasonable (NHBC Foundation, 2008). The Lighthouse (pictured in Plate 8.1) built by Kingspan costs an estimated £180,000, compared to the average £100,000 for a building regulations complaint home (McCarthy, 2008). A slightly lower cost of £150,000 has been estimated for Bill Dunster's Rural Zed house to achieve Level 6 (Building Sustainability, 2008).

Plate 8.1: The Lighthouse at the BRE Innovation Park (Source: BRE)

However, this is a major investment that many home owners are unwilling to make, even with possible savings in running costs of £400 per year. This is a serious concern for housebuilders surveyed by NHBC Foundation, who believe that Code Level 6 homes cannot be built profitably by 2016.

It has even proved challenging for the buildings at BRE's Innovation Park to achieve the standards set out in the code. After some months of further work post-construction, the Lighthouse achieved Level 6 certification. The zero-carbon standard was only achieved after increasing the original target for airtghtness[1] of 1.0m^3/hr/m^2 at 50Pa to 2.30m^3/hr/m^2, and increasing low- and zero-carbon energy supply to meet the extra demand. All of the buildings at the Innovation Park are lightweight stand-alone houses (Yates, 2008), and do not provide solutions appropriate for the urban high-density development that is increasingly encouraged by planning authorities.

Reducing CO$_2$ emissions from new housing

While it is generally agreed that reducing carbon emission from housing is critical, energy use in the home is only part of the picture when it comes to reducing individual CO$_2$ emissions. If less than 20 per cent of our individual carbon emissions are related to energy use in the home, the remainder is due to transport, house materials, shared infrastructure, waste and consumer items, water, services, and around 20 per cent from the production and transport of the food we eat (James and Pooran, 2003).

Energy is used in dwellings for a number of activities. Most commonly, gas is used for space heating and water heating, and electricity is used to power lights, fans, and appliances. Cooking may be by gas or electricity. Statistics show how, over the past 30 years, energy used for water heating and cooking has reduced. However, energy used for space heating has increased despite extensive efficiency measures (see also Chapter 10). The energy used for lights and appliances has also increased due to the large number of appliances we now have in our homes (ONS, 2004).

There are two main measures that can be taken in order to reduce carbon emissions associated with energy use in the home. The first, and most important, is to reduce the energy demand to an absolute minimum. The second is to supply the remaining energy required from low or zero carbon sources. Energy used for water heating, cooking, lights and appliances can be reduced through careful specification and modification of occupant behaviour, but is unlikely to be totally eliminated in modern homes. However, with around 60 per cent of energy used for space heating, reducing this will have a significant impact on CO$_2$ emissions.

---

1 Airtightness is a measure of the rate of movement of air through the building fabric, measured in volume (m^3), per hour, per area of envelope (m^2), at a specified pressure (50Pa).

Reducing space heating energy demand

A home must be heated in order to keep the internal temperature at a comfortable level for the occupants. To reduce heating demand to a minimum, heat losses must first be reduced to a minimum. So long as the temperature inside the building is higher than outside, the heat will be lost through the building fabric. Heat is lost in two ways:

(i) Through the building fabric, i.e. walls, roof, floor, windows and doors.
(ii) Through ventilation, namely controlled ventilation through opening windows and ventilation systems, uncontrolled air leakage through doors, chimneys and gaps in the building fabric.

With a fixed internal temperature required, and an external temperature dependent on climatic conditions, the most effective ways to reduce fabric and ventilation heat loss are:

- **To reduce the exposed areas of external fabric and volume** – more compact forms, such as blocks of flats which have only very small areas exposed, whereas large detached houses have much larger areas of fabric through which to lose heat.
- **To reduce the thermal transmittance of materials used for the external building fabric** – materials with a lower conductivity, thicker insulation, avoidance of thermal bridges (where conductive materials bridge between inside and outside, without being interrupted by insulation).
- **To reduce the rate of ventilation** – the ventilation rate in UK households is commonly uncontrolled, with air leaking through doors, windows and gaps in the fabric. In past years, guidance on airtightness has been largely ignored in the UK. However, the 2006 building regulations introduced an airtightness standard of $10m^3/m^2/hr$, which the construction industry is already finding challenging. Careful detail design and on-site co-ordination between trades can ensure that gaps in the building fabric are reduced.

Building standards

In order to reduce heat loss in new buildings, UK building regulations set limits to U-values[2] for each building fabric element, as well as the airtightness of the building.

As Table 8.1 overleaf shows, building regulations that presently prove challenging for the construction industry can be improved upon further. The requirement of an airtightness value of 10 is in fact not very demanding and has been exceeded for some time in many Central and Northern European countries. The Association for

2 U-values are a measure of the rate of heat loss through the external fabric of the building, in Watts per m^2 area of envelope, per degree difference in the internal and external temperatures. U-values are dependent on the thermal conductivity of the building materials that make up the building envelope.

Table 8.1: Building fabric standards

Building fabric STANDARDS	U-values (Watts per m²K)				Airtightness	
	Roof	Walls	Floors	Windows		
England & Wales (2006)	0.25	0.35	0.25	2.20	10	m³/m²/hr
AECB silver standard	0.15	0.25	0.20	1.50	3	m³/m²/hr
AECB gold standard	0.15	0.15	0.15	0.80	0.75	m³/m²/hr
Passive house	0.10	0.10	0.10	0.80	0.6	AC/hr

Source: Environmental Change Institute: *40% House Report* (2005) – background material F, and David Olivier: *AECB Energy Standards* (8.2.06) prescriptive version p.5.

Environment Conscious Building (AECB) advocates a 'Carbon Lite' standard that aims to exceed 2006 building regulations, and reduce carbon emissions to a minimum. These are the standards that need to be targeted, in order to reduce heat loss sufficiently to bring carbon emissions towards zero. So-called 'passive houses' aim to achieve even higher standards of airtightness (see Table 8.1).

The conventional way of measuring the performance of the building fabric is through U-values. However U-values are quite a crude way of measuring insulation as they are assessed assuming a steady state. In practice, buildings are rarely in a steady state and the external fabric varies according to moisture levels and a variety of other factors. Insulation materials perform differently depending on their nature and the way they are incorporated in buildings. Building regulations require the thermal performance of houses to be predicted in order to comply but it is generally accepted that few buildings actually perform as predicted in practice. Under the European Performance of Buildings Directive (EPBD), buildings are required to be checked, once completed. So-called Standard Assessment Procedure (SAP)[3] and Simplified Building Energy Model (SBEM)[4] ratings are methods used to predict performance but there is little evidence yet as to whether standards are being achieved.

Thermal performance is particularly affected by methods of ventilation (see below), and the design of windows and doors, as well as the layout of the house. There is little point in spending a lot of money on super insulation if cheap windows leak away all the heat. High performance windows and doors with good draught stripping and innovative glazing can make a big difference but come at a price.

Thermal mass

Another issue, which affects the thermal performance of buildings, is the mass of the structure. Lightweight buildings may heat up quickly, but they can also cool down

---

3 Methodology for calculating the energy performance of dwellings.
4 Assessment methodology for carrying out compliance with Part L of the building regulations 2006.

quickly. The concrete and masonry industries are highly critical of timber frame housing for instance, because they say that timber frame buildings are too light and contain no thermal mass. Of course timber frame buildings can also have thermal mass depending on the wall finishes and insulation materials used, but thermal mass can have a beneficial effect in evening out temperature fluctuations and making buildings feel more comfortable. It is also important to remember that in hot summers, even in the UK, houses can become too hot and that insulation and thermal mass are critical to keep them cool as well as warm. This is a crucial point, since climatic changes are now inevitable, with climate change adaptation measures as important as mitigation measures.

Ventilation

It is necessary to provide a sufficient level of ventilation within a home to ensure that levels of pollutants do not adversely affect the health of the occupants. Natural ventilation can be provided by using forces such as the buoyancy of warm air, pressure differences within the building, and wind pressures outside the building, to direct air through the building. Mechanical ventilation utilises fans and ducts to introduce, direct and extract the air in a building. This method usually uses electricity, and will therefore be more expensive to run and crucially produces additional CO_2 emissions. According to the Air Infiltration and Ventilation Centre, 12EJ of the 28EJ of energy consumed by residential buildings, annually in OECD countries, is associated with ventilation. They estimate that this figure could be reduced to as low as one EJ (Concannon, 2002). Part of this reduction can be achieved with the use of ventilation heat recovery. Mechanical systems with improved efficiencies, driven by low wattage fans can produce better CO_2 savings than early models. However, any additional electricity demand will need to be supplied from a renewable energy source, if the target is zero-carbon. Therefore, naturally driven systems are preferable. In order to avoid or reduce energy use, heat recovery ventilation systems can be driven by wind pressure, such as the wind cowls at BedZed, or the stack effect used by 'Pasivent' systems.

Part L and future changes

In the UK, all new homes must meet minimum national standards set by building regulations. The 2006 *Approved Document L1A: Conservation of fuel and power in new dwellings*, sets the required standards in terms of carbon emissions (ODPM, 2006). The SAP calculation estimates the carbon dioxide associated with the energy that will be used for space heating, water heating, lighting, pumps and fans. It does not presently cover energy used for cooling, cooking or appliances.

In order to achieve the government's zero-carbon target, changes to planning and a series of improvements to Part L requirements have been proposed. The document *Building a Greener Future: Towards Zero Carbon Development* (CLG, 2006) details the strategy and a number of measures. These include:

- **Planning** – the *Draft Planning Policy Statement: Planning and Climate Change* sets out how location, siting and design of new developments can deliver the zero-carbon target (see Chapter 5 for a detailed discussion).
- **The Code for Sustainable Homes** – at each of the six levels of the code there are minimum energy/carbon emissions and water efficiency standards. From May 2008, all new homes are required to have a mandatory rating against the code (CLG, 2007d).
- **Building regulations** – in order to achieve zero-carbon homes by 2016, a timetable for progressively improving building regulations over time has been set out:
  - it is proposed that by 2010 new housing achieves a 25 per cent energy/carbon improvement relative to 2006 building regulations, which would be equivalent to Level 3 of the code;
  - this would be increased to a 44 per cent improvement in 2013 (Level 4);
  - leading to zero-carbon housing by 2016 (Level 6) (Sweett, 2007).

Chart 8.1 puts this target into context. According to research by the Association of Environment Conscious Builders (AECB), our present housing stock is responsible for an average of around 72kg CO_2 per m^2 per year (AECB, 2006).

Chart 8.1: Predicted and actual CO_2 emissions reductions leading to zero-carbon homes by 2016

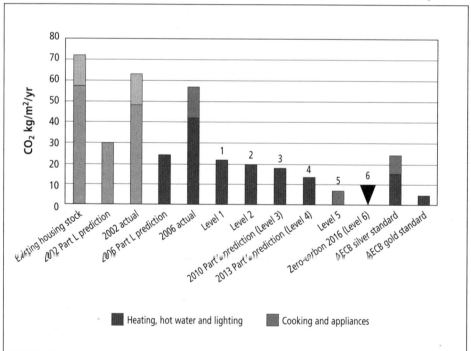

Source: Chart based on data in AECB: *Minimising CO_2 Emissions from New Homes: a review of how we predict and measure energy use in home* (2nd edition) – May 2006. Fig. 2, p.10.

According to this research, the actual CO_2 associated with the operation of new housing could be more than twice as much as that predicted by SAP calculations. There are several reasons for this. The first is that SAP is not an effective modelling tool, and for various reasons does not predict energy demand accurately. Secondly, due to a lack of co-ordination between trades *and* lack of skills within the housebuilding industry, the finished buildings are not as airtight and thermally efficient as they should be. Finally, the occupants may not use the buildings as efficiently as the models assume they do.

When the government set the zero-carbon target, they are referring to zero-carbon in terms of all energy use. Level 5 of the code represents zero-carbon associated with the energy demand of heating, hot water and lighting predicted by SAP. Level 6 represents zero-carbon including cooking and appliances also.

This means that instead of the gradual reduction from 2006 Part L compliant levels to zero, the reductions are likely to be much more of a challenge. In order to achieve zero-carbon homes by 2016, the construction industry will not only have to design and specify thermally efficient homes supplied with low- and zero-carbon energy, but also to ensure they are delivered and operated to perform to this standard. Compliance with and enforcement of these standards is an area where government needs to demonstrate more commitment. A study by the Energy Efficiency Partnership for Homes amongst building control officers on the compliance with Part L1 of the 2002 building regulations revealed that on the whole officers will not enforce or refuse completion certificates or prosecute on a Part L issue (EEPH, 2006). Part L was given low priority as it was seen as not being 'life threatening'. This practice clearly has to change and in order to achieve the zero-carbon target, energy-efficiency requirements need to be enforced rigorously. There is also some evidence that development control officers are still getting to grips with the 2006 changes to Part L (Lazell *et al.*, 2007).

The Code for Sustainable Homes

The Code for Sustainable Homes is a new assessment method introduced in 2007, which, as well as defining energy standards, also sets standards for improvements in the overall environmental performance of new homes. Based on the EcoHomes method, which it replaces, the code awards improved performance in terms of nine environmental issues. Under the code, a home is assessed at design and post-construction stages, before it can be certified with a rating of Level 1 to 6, Level 6 being the highest zero-carbon standard (CLG, 2007e). All new housing is required to have a code certificate in the Home Information Pack. Whilst it does not presently require an assessment, or a specific level of performance for private housing, it does require a mandatory rating. However, the government requires that all new publicly-funded housing meets Level 3 of the code. The Housing Corporation (soon to merge with English Partnerships to become the Homes and Communities Agency) already requires Level 3 for compliance with Scheme Development Standards for the current bidding round 2008-2010. This requirement is likely to increase to Level 4 by 2010, and Level 6 by 2013. Planning authorities are increasingly setting code levels as requirements for private developments (see Chapter 5).

Construction techniques

Modern Methods of Construction

Modern Methods of Construction (MMC) have the potential to help meet the challenges associated with the delivery of zero-carbon homes in several ways:

- factory-based, trained and skilled workforce producing better quality;
- co-ordinated construction process; and
- predictable thermal performance.

The Housing Corporation classify MMC in terms of five categories (NHBC Foundation, 2006).

1. **Off-site manufactured – Volumetric:** These are complete three-dimensional units produced in a factory, offering the potential for effective airtightness, through controlled workmanship (EST, 2005). However, these units arrive almost complete and are limited in terms of future adaptability.
2. **Off-site manufactured – Panelised:** The three-dimensional structure is assembled on-site, using flat panel units produced in a factory. Providing they are well-constructed these panels do also offer opportunities for improved airtightness, reduced thermal bridging and improved thermal performance, particularly around openings such as windows. However, structural panels also offer less opportunity for future adaptability compared with framed structures.
3. **Off-site manufactured – Hybrid:** Combining volumetric elements such as bathrooms and kitchens with panelised elements such as walls to create the homes. Although there is again the potential for improved thermal performance, great care must be taken in the detailing, particularly where elements come together, as thermal bridging and airtightness can be affected.
4. **Off-site manufactured – Sub-assemblies and components:** Use of factory-made components rather than complete systems, such as roof and floor cassettes, pre fabricated foundations and pre fabricated roof structures and dormers. Again, care must be taken with the interfaces between different components.
5. **Non off-site manufactured modern methods of construction:** Site-based construction methods that are more innovative than traditional methods. These might use in-situ concrete methods such as insulating formwork, tunnel form construction or block and plank methods. The in-situ methods have the potential to be very airtight and provide structures of high thermal mass.

Recent research (Barett and Wiedmann, 2007) has indicated that off-site manufactured housing has the potential to reduce carbon dioxide emissions by 30 per cent compared with on-site construction. However, the evidence also indicates that off-site manufactured housing may offer less potential for future adaptation and the integration of new technologies. The houses built at BRE's Innovation Park using MMC have had difficulties achieving the high airtightness standards needed to meet code levels. It is thought that this is due to the challenge of junctions between different panels and components (Yates, 2008).

In a recent article, Sir Digby Jones, former Director General of the CBI, raised concerns regarding future maintenance costs for homes built using MMC (Jones, 2007). He remarked that pods, components and panels:

'...are often made using low-cost labour in conditions that would not be acceptable in the UK, are transported thousands of miles and are erected using minimal site labour – often brought in by the installer, so providing no local employment. They look nice and shiny, but what will these homes be like to live in in 20 years' time?' (ibid.).

He warns that the result will be housing with a short life, that is expensive to maintain and uncomfortable to live in. He concludes with a suggested improvement on the definition of the target for zero-carbon housing by 2016:

'...to make all new-build dwellings carbon-neutral from the first stage of construction to the last stage of demolition, with a target life of at least 200 years and to be able to sustain a 4°C rise in summer temperatures over that period' (ibid.).

Energy supply

Once homes have reduced energy demand to a minimum, the remainder of the energy must be provided from a low- or zero-carbon source.

Since the early 1990s, Woking Borough Council has been developing local energy systems. Money saved through energy-efficiency projects was re-invested in expanding local energy capacity, presently totalling over 60 local generators. These include cogeneration, trigeneration, photovoltaic arrays and the UK's first hydrogen fuel cell station (Trade and Industry Select Committee, 2007). As a result, the council is said to have reduced its CO_2 emissions by 77 per cent, and the Borough of Woking's emissions by 17 per cent.

Building on the experience at Woking, the Greater London Authority (GLA) aims to achieve a 20 per cent reduction on 1990 emission levels by 2010. In order to do this, new developments are required to reduce their CO_2 emissions by at least ten per cent by supplying energy from on-site renewables. This requirement is expected to increase to 20 per cent. Developments are also expected to consider Combined Heat and Power (CHP) and heat-fired absorption cooling. The involvement of Energy Service Companies (ESCo's) is expected to help develop local energy networks, in the same way that Woking has established a private wire network to distribute local energy. However, the success in Woking has yet to be replicated. Attempts by the London Development Agency to create a demonstration of urban wind power potential through the installation of 14 wind turbines on the roof of the Pallestra building, resulted in minimal output, and their removal after several months.

Merton Borough Council responded to the 2004 *Planning Policy Statement on renewable energy* (PPS 22) by setting a target that all non-residential developments above 1,000 m^2 supply at least ten per cent of their energy from renewable technologies. By 2007, 80 other local authorities had followed this example. Unfortunately, the recent *Planning Policy Statement on climate change* (CLG, 2007), failed to set similar targets nationally. There are concerns from the building industry regarding the extra cost in meeting the ten per cent requirement. There are also serious concerns about the impact particularly on urban developments. It is thought that a drive to meet this target will result in a reduction in the quality of buildings, and the specification of inappropriate energy solutions that may not perform as predicted (Butler, 2007). Government has responded by saying that off-site renewables could also be allowed (Seager, 2007), and a private members bill (Fallon, 2007) supporting the Merton Rule has been introduced to encourage micro-generation across the UK.

Concerns have also been raised regarding the heavy reliance on biofuels in order to meet both the Merton Rule and the zero-carbon targets. DEFRA's chief scientific adviser Bob Watson has warned that biofuels could '...*exacerbate climate change rather than combat it'* (Renderson and Watt, 2008). It seems logical that if we have a method for storing carbon, we should be building *with* it, not burning it (see Chapter 9 for a comprehensive discussion).

Examples of practice

It is a relatively straightforward exercise to design low-carbon homes in theory, on paper, but how do they perform in reality? Many Central and Northern European countries have been designing and operating housing to higher energy-efficiency standards for some years. There have also been a number of pioneers in this country working to meet the challenge of affordable energy-efficient housing. There is an opportunity to examine the performance of these exemplar developments, to inform present strategies for achieving low- and zero-carbon housing in the UK (for a detailed discussion on the impact of pioneering schemes on national policy see Chapter 7).

Task 13 houses

The International Energy Agency's Solar Heating and Cooling Programme examined the performance of passive and active solar technologies, and advanced energy-efficiency measures for reducing heating and cooling loads in residential buildings. The average energy used for space heating by the IEA's 'Task 13' houses (Hestnes *et al.*, 1996) was reduced to 14kWh/m^2, compared with 97kWh/m^2 required for a typical house. This was achieved by specifying very low U-values, airtight building fabric, and by taking advantage of solar and internal heat gains. The houses were also supplied with controlled ventilation and heat recovery. An analysis of the heating load for each house reveals that an average of around 24 per cent of space heating requirements are met through heat recovery. However, the 'Task 13' houses use an average of 6kWh/m^2/yr of electricity to run the fans and pumps of the ventilation heat recovery systems, compared with the 19kWh/m^2/yr of heating energy saved due to the use of heat recovery. Electricity generated by burning fossil fuels in power stations and distributed through the grid, is responsible for more than twice the CO_2 associated with burning gas to heat a home. If the electricity to run the fans and pumps is supplied from the grid, the associated CO_2 emissions are almost equal to those saved by displacing the gas used to heat the home. Therefore, there was almost no carbon dioxide benefit in using heat recovery for these homes. This could have been further worsened by reductions in fan efficiency, due to age or lack of maintenance. This could make the heat recovery solution even more carbon intensive than gas heating. Higher efficiency heat recovery and lower wattage fans, developed recently, promise better carbon performance. However, this example does highlight the dangers when applying technology to solve a problem, namely that there often emerges a new problem, created by the solution.

Passive house

Passive house standards applied in Central and Northern European countries have achieved reductions in space heating energy demand down to as little as one eighth of a conventional house (Passive House Institute, 2007). Space heating requirements are as low as 10 to 20kWh/m^2/yr. This is achieved through the application of the following:

- insulation;
- design without thermal bridges;
- maximum airtightness;
- ventilation with heat recovery;
- triple-glazed low-e windows with insulated frames; and
- innovative heating technology.

Research into 150 passive houses in the German Land of Nordrhein-Westfalen (NRW) was carried out in order to evaluate their performance, and the satisfaction of the occupants painted a mixed picture (Berngen-Kaiser and Frey, 2006). Around 54 per cent of the houses experienced problems with the mechanical heat recovery ventilation systems. In 14 per cent of the houses, the equipment still does not work properly. These problems were mainly due to the commissioning of the systems, lack of skills among the installers and some errors in the design of the systems. Problems with noise from the systems were experienced by 56 per cent of residents, while 39 per cent reported that they did not hear any noise. While most residents were comfortable in the winter months, 37 per cent of residents stated that temperatures were often too high in the summer. In terms of energy performance, the results were disappointing, with less than half the homes keeping within the limiting value for primary energy of 120kWh/m^2/yr. The research highlighted some interesting issues:

- energy use in homes where the occupants were motivated by personal conviction to build a passive house, was lower than those who were not;
- the high levels of comfort experienced in the winter may not have helped in motivating the occupants to monitor or save energy;
- 87 per cent of the occupants had been given no instructions on the operation of their home; and
- the lack of skills and commissioning produced serious performance problems with the ventilation systems.

These results provide important lessons for the development of zero-carbon standards in this country, namely, that:

(i) technologies must be installed correctly using a skilled workforce, monitored, maintained, and adjusted as needed to maximise performance;
(ii) occupants must be motivated to operate their homes efficiently and be given sufficient information and training to do so. Simple low-tech solutions may not perform as well in predictions, but may perform better than high-tech solutions in terms of actual operation.

Enhanced Energy Performance Standard

The Partners in Innovation Project recently completed research into the effectiveness of the Enhanced Energy Performance Standard (EPS08 Energy Standard), by assessing a development of 700 cavity masonry dwellings at Stamford Brook (Wingfield et al., 2007). Despite significant improvements over 2002 building regulations in force at the time of construction, important differences between the designed (25-35 per cent improvement over 2002 building regulations), and actual energy performance of the houses were identified:

- 37 per cent of the 44 dwellings tested did not achieve the airtightness target of 5m^3/(h.m^2) @ 50Pa;
- one of the most significant findings was that the whole house heat loss coefficients were in some cases 100 per cent higher than predicted, and that

the major cause of this was heat lost through the party-wall cavity, previously assumed to be insignificant;
- differences between modelled and actual U-values were also identified, meaning that more heat was lost through the building fabric than had been predicted (the reasons for this are thought to be related to a lack of skill and knowledge in the construction industry regarding both detail design and construction that is needed for improved thermal performance); and
- due to inefficient heating system design plus programming errors, measured boiler efficiencies were found to be 2-6 per cent below their SEDBUK[5] rating of 91.3 per cent, and overall system efficiencies as low as 50 per cent in the summer.

Most significantly, the authors identify the major issues as relating to the '...*interrelationship of the various parts of the construction process from design conception all the way through to completion and occupation*'. This indicates that the way that the construction industry operates will have to change fundamentally, particularly with low- and zero-carbon housing planned for the near future.

The research also indicates that Code Level 3 energy performance is achievable using similar, but slightly enhanced standards applied at Stamford Brook. However, Level 4 without low- or zero-carbon technologies could only be achieved by applying passive house standards plus whole-house heat recovery ventilation solutions.

Conclusion

It is clear that achieving zero-carbon homes is not an easy task. While it can be achieved in theory, there is a long way to go in terms of the practical implementation. The most important lesson, it seems, is the design of simple passive low-energy homes, and the appropriate application of solutions, particularly technologies. Ambitious targets for housebuilding, are driving the construction industry towards quick low-cost solutions, that tick the boxes set by government. There is a danger that these solutions will increase environmental impact not decrease it, if they are not applied in an informed manner.

Many Central and Northern European countries have years of experience of designing and building low-energy housing. These countries are not targeting zero-carbon. Is this because 'zero' is not seen as a realistic target? They have developed the knowledge and skills to deliver housing effectively with reduced energy demand equivalent to Level 4 of the code. In order to improve performance to Levels 5 and 6, it will be necessary to rely on micro-generation, specifically biomass Combined Heat and Power. The zero-carbon status of housing will be dependant on the performance of these technologies, which in turn relies on appropriate design, effective

5 Seasonal Efficiency of a Domestic Boiler in the UK.

installation, commissioning and maintenance. Initially, it may be more useful to set a target of Level 4, and ensure that performance matches design predictions. Otherwise, we risk diverting funds away from the delivery of quality housing, to micro-generation in order to achieve Level 6 housing; housing which may in the long term struggle to perform past Level 3 standards. This concern is echoed by recent research (Williams, 2008). Williams concludes that '...the housing programme will need to be slowed' to enable the capacity for delivery of zero-carbon homes and lifestyles to be developed. She suggests that what is likely to be most effective is '...a combination of passive technologies maintained and managed by external service providers in new housing, and centralised zero-carbon energy supply'.

In conclusion, there are a number of factors that are likely to prevent the delivery of zero-carbon homes by 2016, and some important questions that need to be addressed:

- Is zero-carbon by 2016 a realistic and useful target for reducing domestic carbon dioxide emissions effectively?
- Should we include embodied carbon, and lifestyle carbon within the definition of zero-carbon homes?
- If the most urgent environmental target is zero-carbon, then why is it tied up with other environmental (and some social targets) in the code, since challenging water and materials targets are preventing some developers, particularly in urban areas, from targeting Level 4?
- Is there sufficient understanding of measures appropriate for high-density developments to achieve zero-carbon?
- Does the construction industry have the skills and commitment to deliver zero-carbon homes?
- Will the technology be maintained at levels necessary for housing to perform as zero-carbon in the long term?
- Do we really want to drive every new housing development to specify biofuels?

References

AECB (2006) *Minimising CO2 Emissions from New Homes: a review of how we predict and measure energy use in homes*, Sustainable Building Association. Available from: http://www.cseng.org.uk/_db/_documents/Minimising_CO2_Emissions_from_New_Homes_-_AECB.pdf

Barrett, J. and Wiedmann, T. (2007) *A Comparative Carbon Footprint Analysis of On-Site Construction and Off-Site Manufactured House*, Stockholm, Environment Institute and Centre for Integrated Sustainability Analysis. Available from: http://www.isa-research.co.uk/docs/SEI_ISA-UK_Report_07-04_OSM_House.pdf

Berngen-Kaiser, A. and Frey, T. (2006) *Evaluation of 150 Passive Houses in NRW*: Paper included in Conference Proceedings of 10th International Passive House Conference 2006

Boardman, B. (2007) *Home Truths: A low-carbon strategy to reduce UK housing emissions by 80% by 2050*, Environmental Change Institute, Oxford University. Available from: http://www.eci.ox.ac.uk/research/energy/hometruths.php

Brinkley, M. (2008) *One year on, is the Code for Sustainable Homes working?*, Communities and Local Government, Building Design, 20 March. Available from: http://www.bdonline.co.uk/story.asp?sectioncode=427&storycode=3109333&c=2&en cCode=00000000014886a0

Building Sustainability (2008) 'Rural Zed Explained', *Building*, Issue 12. Available from: http://www.building.co.uk/sustain_story.asp?sectioncode=331&storycode=3109716& c=2

Butler, T. (2007) 'The rules that stop us going for the burn', *Building Design*, 5 October. Available from: http://www.bdonline.co.uk/story.asp?sectioncode=453& storycode=3096846

CLG (2006) *Building a Greener Future: Towards Zero Carbon Development*, Communities and Local Government. Available from: http://www.communities.gov. uk/archived/publications/planningandbuilding/buildinggreener

CLG (2007a) *Homes for the future: more affordable, more sustainable –* *Housing Green Paper*, Communities and Local Government. Available from: http://www.communities.gov.uk/publications/housing/homesforfuture

CLG (2007b) *Zero Carbon Homes*, Communities and Local Government. Available from: http://www.communities.gov.uk/speeches/corporate/zero-carbon-homes

CLG (2007c) *Eco-towns Prospectus*, Communities and Local Government. Available from: http://www.communities.gov.uk/publications/housing/ecotownsprospectus

CLG (2007d) *The Future of the Code for Sustainable Homes: Making a rating mandatory,* Communities and Local Government. Available from: http://www.communities.gov.uk/publications/planningandbuilding/futurecode consultation

CLG (2007e) *Code for Sustainable Homes: Technical guide*, Communities and Local Government. Available from: http://www.planningportal.gov.uk/uploads/code_for_ sustainable_homes_techguide.pdf

CLG (2007f) *Planning Policy Statement: Planning and Climate Change*, Communities and Local Government. Available from: http://www.communities.gov.uk/publications/ planningandbuilding/ppsclimatechange

CLG and EP (2006) *Designed for Manufacture: The challenge to build a quality home for £60k – Lessons learnt*, Communities and Local Government, English Partnerships. Available from: http://www.designformanufacture.info/lessonslearnt.htm

CLG and EP (2007) *Carbon Challenge standard brief: The challenge to build quality sustainable homes*, Communities and Local Government, English Partnerships. Available from: http://www.englishpartnerships.co.uk/carbonchallenge

Concannon, P. (2002) *Residential Ventilation Air*, Infiltration and Ventilation Centre Technical Report TN 57. Available from: http://www.aivc.org/frameset/frameset. html?../publications/publications.html~mainFrame

Concrete Centre (2006) *£60K house will have to be paid for by increased carbon dioxide emissions*, The Concrete Centre. Available from: http://www.concretecentre. com/main.asp?page=1235

EEPH (2006) *Compliance with Part L1 of the 2002 Building Regulations: an investigation on the reasons for poor compliance*, Energy Efficiency Partnership for Homes. Available from: http://www.eeph.org.uk/resource/partnership/index.cfm? mode=view&category_id=35

EP (2007) *Millennium Communities Programme*, English Partnerships. Available from: http://www.englishpartnerships.co.uk/millcomms.htm

EST (2005) *Building Energy Efficient buildings using modern methods of construction*, Energy Saving Trust. Available from: http://www.energysavingtrust.org.uk/uploads/ documents/housingbuildings/CE139%20-%20modern%20methods%20of%20 construction.pdf

Fallon, M. (2007) *Planning and Energy Bill 2007-08*. Available from: http://services.parliament.uk/bills/2007-08/planningandenergy.html

Hestnes, A. G., Hastings, S. R. and Saxhof, B. (1996) *Solar Energy Houses: Strategies, Technologies, Examples*, London: James & James Science Publishers

House Builders Federation (2007) *Zero Carbon Homes – Delivering the 2016 target*, House Builders Federation

James, P. and Pooran, D. (2003) *One Planet Living in the Thames Gateway*, World Wildlife Fund. Available from: http://www.wwf.org.uk/sustainablehomes/reports.asp

Jones, D. (2007) 'Out with the New', *Building,* Issue 8. Available from: http://www.building.co.uk/story.asp?sectioncode=31&storycode=3081688

Lazell, M., Ancell, H., Bennett, E. (2007) '"We're not ready" for zero carbon rules', *Building Design*, 16 February. Available from: http://www.bdonline.co.uk/story.asp? sectioncode=1100&storycode=3001530&c=20cncCode=0000000001200ff7

McCarthty, M. (2008) 'On the Market: the zero carbon home with an affordable price', *The Independent*, 27 February. Available from: http://www.independent.co.uk/ environment/green-living/on-the-market-the-zerocarbon-home-with-an-affordable-price-787920.html

NHBC Foundation (2006) *A Guide to Modern Methods of Construction*, NHBC Foundation

NHBC Foundation (2008) *Homeowners are not ready for zero carbon homes, research shows*, National House-Building Council. Available from: http://www.nhbc. co.uk/Newscentre/Recentnews/Name,33094,en.html

NHF (2007) *Government in danger of missing target for all new homes to be zero carbon by 2016*. Available from: http://www.housing.org.uk/default.aspx?tabid=212& mid=828&ctl=Details&ArticleID=654

ODPM (2006) *Approved Document L1A*, Office of the Deputy Prime Minister. Available from: http://www.planningportal.gov.uk/england/professionals/en/ 1115314231792.html

ONS (2004) *Social Trends 34*, Office for National Statistics. Available from: http://www.statistics.gov.uk/StatBase/Product.asp?vlnk=5748

OPSI (2007) *Stamp Duty Land Tax (Zero Carbon Homes Relief) Regulations 2007*, Office of Public Sector Information. Available from: http://www.opsi.gov.uk/si/ si2007/uksi_20073437_en_1

Passive House Institute (2007) *Information on Passive House*. Available from: http://www.passivhaustagung.de/Passive_House_E/passivehouse.html

Renderson, J. and Watt, N. (2008) 'Top scientists warn against rush to biofuel', *The Guardian*, 25 March. Available from: http://www.guardian.co.uk/environment/2008/ mar/25/biofuels.energy1

Seager, A. (2007) 'Parties unite to stop government backsliding over renewables rule', *The Guardian*, 3 December. Available from: http://www.guardian.co.uk/ business/2007/dec/03/energy.renewableenergy

Shaw, R. (2007) *Eco-towns and the next 60 years of planning*, Town & Country Planning Association. Available from: http://www.tcpa.org.uk/press_files/ pressreleases_2007/20070924_TS.pdf

Smith, A. (2007) *Zero Carbon Homes*, Communities and Local Government. Available from: http://www.communities.gov.uk/speeches/corporate/zero-carbon-homes

Sweett, C. (2007) *A cost review of the code for sustainable homes: Report for English Partnerships and the Housing Corporation*, English Partnerships, Housing Corporation. Available from: http://www.cyrilsweett.com/pdfs/Code%20for%20 sustainable%20homes%20cost%20analysis.pdf

Trade and Industry Select Committee (2007) *Local Energy – Turning Consumers into Producers: First Report*, House of Commons. Available from: http://www.publications. parliament.uk/pa/cm200607/cmselect/cmtrdind/257/25702.htm

Williams, J. (2008) 'Green Houses for the Growth Region', *Journal of Environmental Planning and Management,* Vol. 51, No. 1, pp.107-140

Wingfield, J., Bell, M., Miles-Shenton, D., Lowe, B. and South, T. (2007) *Stamford Brook: Evaluating the Impact of an Enhanced Energy Performance Standard on Load-bearing Masonry Domestic Construction*, Leeds Metropolitan University

WWF (2008) *How Low: Achieving optimal carbon savings from the UK's existing housing stock*. Available from: http://www.wwf.org.uk/filelibrary/pdf/how_low_report.pdf

Yates, A. (2008) Conversation with attendees of Tall Buildings meeting at BRE Innovation Park dated March 31st 2008

CHAPTER 9:
Low impact methods of construction – the way forward?

Tom Woolley

Introduction

Many people believe that we can find our way out of global warming and other environmental problems through technological solutions. However, there are good and bad technologies and the construction industry, in particular, seems far too easily drawn to what appear to be quick fix technological solutions, particularly if they are sold as more efficient and using less labour. In this chapter I will consider the role that natural and low impact alternative materials can play in sustainable construction of housing. The now retired Chief Scientific adviser to the UK government, Sir David King, advanced the reassuring message that science and technology will come up with the answers, *'…we must redouble our efforts to develop the technologies we need to achieve a low carbon economy'* (King, 2007).

Unfortunately this leads many people in Western developed nations to assume that we can maintain our existing lifestyles without too much difficulty and that radical changes in what we do are not really necessary. The word 'technology' sounds like it means continuing with fossil fuel powered cars and trucks, synthetic, petro-chemical based building materials and cheap air travel. Such is the pre-occupation with defending the status quo, which is deeply embedded in most conservative societies, that there is enormous resistance to accepting alternative and innovative solutions to our environmental problems.

Those of us who advocate alternative low impact and low-carbon solutions to building houses come up against a great deal of what I can call 'techno-prejudice' because what we are offering looks like turning the clock back. Kochan (2008) for instance refers to hemp and lime construction as a 'traditional material'. For a method of construction that is only ten years old and has required huge investment in finance and scientific research this is a strange description, but because it uses simple materials with minimal processing, it is perceived by some as 'traditional'.

Science and technology does hold many of the answers to how we can live in harmony with the planet, but very little of the right science and technology is being pursued by government and the mainstream housebuilding industry. Natural, ecological building materials and products fall foul of this because investment is largely going into short-term high-tech solutions that are *not* sustainable.

For instance, if we can produce high performance building insulation materials from natural crop-based products like wood, straw, sheep's wool, hemp, flax and even soya and sugar starch, that absorb carbon dioxide, use minimal energy to produce and can be recycled; why not use them in preference to synthetic fossil fuel-based materials? If these natural products are also largely free of toxic chemicals, can create much healthier buildings and can absorb moisture, thus reducing condensation risks, then you would expect architects, housebuilders and the general public to choose them in preference to conventional building materials. Indeed such materials can produce better, longer-lasting buildings that are more robust. To some of us it seems obvious that we should be developing a new sustainable low-carbon economy based on such materials and products and they should be used in housebuilding. But there is still a great deal of prejudice and hostility to these ideas. Growing building materials from crops may seem to be as controversial as recent concerns about biofuels, since taking valuable agricultural land away from food production is not sustainable. However, growing materials like hemp cannot be seen in the same terms as using cereal crops for petroleum. Hemp can be grown on set aside land or can be used to clean up the ground between cereal and other food crops. It actually helps to improve food yields.

Such resistance to using innovative natural materials exists *even* when high-tech solutions can be shown to be failing. The purpose of this chapter is to try and convince sceptical readers that natural ecological materials and products are now readily available, that they work as well, if not better than conventional approaches and that they are scientifically proven and affordable (or could be).

Unless we make a speedy shift to producing and using natural alternative materials, conventional products which rely on a great deal of fossil fuel will soon be unaffordable and unavailable as oil prices rise and environmental pressures increase. Modern buildings have the potential to make people ill and sooner or later litigation will force landlords and developers to turn to healthier solutions. Already in the USA and other developed countries we can see class actions about indoor pollutants and toxic mould. UK insurance companies have introduced exclusions around these issues. Natural ecological ways of building will not just be for a minority who want to enjoy a green lifestyle, but will eventually be a central part of mainstream construction once it is more widely recognised as resolving indoor pollution problems.

As policies and legislation require higher standards of environmental declaration and much lower carbon emissions, mainstream construction will not be able to keep up with more stringent requirements. Vested interests in the increasingly global construction materials market will and are already putting up a stiff resistance to changes and are attempting to water down and resist new more rigorous standards, but this is only putting off the inevitable.

Timeless buildings?

A study carried out for British Gas found that a 16th century half-timbered house was more airtight and energy efficient than many houses constructed since the Second World War (British Gas, 2006).

Table 9.1: House construction and airtightness

Building Type	Leakage
Tudor	10.11
1960s	15.1
1970s	11.7
1980s	12.0 – 40.1
1990s	12.0 – 23.6

Source: British Gas (2006).

British Gas suggested that an air leakage index of 8 was a best practice standard and that 15 was more normal (ibid.). Building regulations require a standard of 10. The smaller the figure, the better the airtightness and thus the energy efficiency. However, hemp and lime and so-called passive house standard buildings can achieve an airtightness of less than 2. If this is achievable, without too much difficulty, then we need to ask why such a poor standard is all that is required to meet the building regulations. The answer is quite simple; it is to allow poorer forms of construction to be approved under the regulations.

Traditional buildings in the past, used natural materials, sometimes from the locality, timber, locally slaked limestone, earth, mud, straw and so on. Following the invention of Portland cement we moved more and more to high-energy intensive brick and block manufacture. Increasingly synthetic materials like upvc, mastic sealants, plastic membranes and so on were introduced. Today we can see the beginnings of a backlash against the plastics era. Wealthy people want hardwood windows rather than PVC. PVC windows are said to knock thousands of pounds off the value of a house today. On the downside, the hardwood may have been illegally logged from a rain forest.

As many people turn towards natural and organic, locally produced foods (if they can afford it), awareness of the value of traditional and natural ways of doing things is growing. Many new natural building products involve age-old materials like earth and lime because it is known that traditional methods have stood the test of time. Oak posts and beams treated with lime have lasted centuries. Walls of mud and straw (cob) have lasted almost as long until ignorant modern builders started rendering them with cement and sand and non-breathable 'waterproof' coatings. Traditional builders understood the importance of buildings being able to breathe.

Lime washes and natural oils could add better weather protection than more recent synthetic plastic and polymer coatings that seal in dampness. Old buildings also used stone or brick, which added thermal mass to buildings, frequently absent from modern buildings.

Builders and developers today will say that we cannot put the clock back and that such traditional ways of building are too expensive, do not work, are too slow and do not meet modern expectations. Even in poorer developing countries, modernisation means abandoning traditional wisdom about building construction and design in favour of western style manufactured materials.

However, it is possible to use natural materials in modern ways in which they can go a long way towards current requirements. These modern formulations of natural materials are no longer 'traditional'. Hemp and lime composite construction, described below, is a modern invention and has been used in restoring historic buildings in place of the old wattle and daub wall infill. There are many people who are expert in restoring old buildings and frequently they run into problems meeting new energy and sustainability standards, but there is also much we can learn from their sensitive understanding of how buildings really work (Ryan, forthcoming).

Creating fake pastiche 'old world style' buildings with modern materials like concrete, steel and plastic is not a sustainable approach. The Prince's Foundation, which has done a lot of good work to reawaken interest in traditional building materials and methods, has sadly also conferred legitimacy on the fakery of Poundbury and other 'Georgian style' developments.

Using natural materials instead of synthetic components

The idea of natural building may call up images of strawbale houses and tepees in the countryside for many people but over the past few years there has been a significant growth of mainstream construction projects adopting natural building materials and methods. There has also been a rapid growth in the manufacture of natural, low impact and crop-based products. Such building methods and products could make a significant contribution to efforts to reduce carbon emissions and resource consumption because of the reduced reliance on energy and petro chemicals. However, the development of a natural building product industry is being hindered by a lack of government support and a failure to realise that low impact construction should be a central feature of zero-carbon and sustainable development. The Department of Trade and Industry has been carrying out a series of reviews of government sustainable construction strategies and the latest consultation version does at least give a passing mention to natural and renewable materials though it may not appear in the final document (BERR, 2007). The Department of

Environment, Food and Rural Affairs (DEFRA) is supporting renewable crop-based materials through the establishment of the National Non Food Crops Centre in York which is promoting the use of renewable materials for various industries including construction (NNFCC, 2008).

Unfortunately government policy tends to base its low-carbon building policies largely on renewable energy and micro-generation, water and waste reduction. While adding renewable energy to buildings (micro-generation) can provide a useful contribution to tackling global warming and reducing energy costs, many are questioning its sustainability. Grants are given for renewable energy without necessarily requiring buildings to be particularly energy efficient. Making a building very energy efficient (with lots of insulation and good airtightness) will have much quicker payback than renewable energy but does not provide such good photo opportunities for politicians. PV panels on a roof are easy to spot whereas insulation, hidden away in the building fabric, is far less 'sexy'!

There are also those who argue that energy efficiency should best be achieved using synthetic insulation materials like glass fibre, polystyrene and urethane foams even though a lot of fossil fuel energy is required to produce such materials. They tend to scorn natural materials as too expensive and dismiss issues of embodied energy and the health and pollution impacts of synthetic materials. With such opponents, advocates of natural building have a tough job convincing society of its advantages though surprisingly the market for natural materials is growing rapidly as many people are willing to use them, even when they cost more.

Arguments for and against natural materials and building methods are as follows.

The Pros

- Natural materials use much less energy to produce and have significantly less damaging effects on the environment. The gains from using natural materials are almost instant and thus if materials like unfired earth and crop-based materials were used, pollution, fossil fuel use and toxicity would be reduced in a very short period of time. The energy-efficiency gains of using synthetic materials in 10 or 20 years cannot be justified, as we need to reduce fossil fuel use and emissions right away not in 20 years time.

- Natural materials provide many additional benefits that are not available from synthetic materials. They can retain heat and create a thermal lag effect that evens out temperature fluctuations. This is found in the thermal mass of unfired earth or even lighter weight insulation materials. Natural insulation and walling materials usually perform better than might be predicted from conventional 'U-value' calculations because conventional building science has not taken into account the different properties of natural materials.

- Natural materials can be significantly beneficial in terms of indoor air quality because they can absorb moisture and regulate humidity, reducing risks of toxic mould growth and condensation, allowing moisture to permeate through the building fabric. Natural materials have much lower levels or no toxic additives such as flame-retardants that can affect health. Indoor air quality (IAQ) is not a big issue in the UK unlike the USA and other European countries but minimum IAQ standards can now be found in UK building regulations (Yu and Crump, 2007). Toxic mould and Volatile Organic Compound (VOC) emissions are a serious health hazard but these can be less of a problem if natural materials are used.

- Natural materials create significantly less pollution to the wider environment both in production and at end of life whereas synthetic materials use toxic flame-retardants, glues and binders. Natural materials decay back into the ground without any harmful effects. They are safer to handle and reduce health risks for builders and building occupants.

- Natural materials are fundamentally better in terms of sustainability. Traditional methods such as cob (mud and straw), lime renders and solid timber have lasted for hundreds of years. 'Quick fix', high-tech building methods which are being rushed out to meet the demand for new housing are already failing after a few years. We are in danger of seeing another decade of system building disasters as happened in the 60s and 70s and already some innovative volumetric projects are being demolished, after only a few years.

The Cons

- Many natural materials are expensive because they are largely imported from outside the UK. Manufacture of typical natural products from wood fibre and hemp takes place mainly in Poland, Austria, France, Germany and Switzerland. These materials could easily be made in the UK and the raw materials are locally available. Ironically some of the hemp insulation products imported from mainland Europe use hemp grown in the UK! As the raw materials are not inherently expensive it is largely a question of scale of production. The more we use the cheaper they will be. Also in Germany, natural insulations are subsidised through government grants to householders who install them. Builders and quantity surveyors often overprice natural materials and methods when preparing estimates, because they are unfamiliar and innovative. Availability is also a problem, as natural materials are not as easy to source since too few merchants stock the products.

- There are strong prejudices against using natural materials as there is a misplaced belief that if something is toxic, plastic and synthetic, it must be better and will last longer! The myth has been created that upvc windows are low maintenance and thus to be preferred to timber windows. However, many PVC windows installed in the 80s and early 90s are already due for replacement.

- Natural methods of construction can involve wet trades and may be slower to dry out. This seems like turning the clock back to an industry encouraged to use dry systems. Some natural methods depend on site assembly. However, many products and materials can be fabricated off-site and simply substituted for synthetic equivalents.

Examples of natural building products and methods

Unfired earth:
Bricks made from clay are usually fired at high temperatures but there is a revival of unfired earth methods of construction. Cob is the traditional method, a handcraft technique using mud and straw. There has been a renewed interest in cob, not only for the restoration of old buildings, but also for new development. Rammed earth is often used in large building projects but infrequently in housing. Major brick and block manufacturers have now developed new unfired earth bricks and block products, which are available through regular builders merchants.

Strawbale:
Strawbale building is a labour intensive handcraft technique, attractive to self-builders but is also available in prefabricated panels, assembled off-site. It is unlikely that strawbales would be used in urban or social housing developments but prefabricated panels could be used in these situations.

Plate 9.1: Strawbale self-built private house, Putley, Herefordshire (Photo: Tom Woolley)

Hemp and wood fibreboards:

A growing range of building and insulation boards made from hemp and wood fibres, often using natural resins or very low toxicity glues, are being manufactured. Many of these boards are impregnated with natural waterproofing materials such as latex to make them more weather resistant. Hemp loft boards can be bought at DIY stores and builders merchants. Wood-fibre sarking boards have been extensively used in social housing projects in Scotland and England. They provide an attractive option for well-insulated warm roofs. Some of these boards can be used as external insulation in renovation projects.

Hemp insulation:

Hemp insulation batts and quilts are now so popular that it has been hard for the manufacturers to keep pace with demand. 'Isonat' and 'Breathe' products are made from hemp, recycled cotton and a polyester binder. Hemp used in insulation and Hemcrete (see below) can be grown in almost any part of the UK. One hectare of arable farmland can produce about ten tonnes of hemp which is enough to build a typical small house using Hemcrete. The straw, which is about 80 per cent of this weight, is used for Hemcrete and the fibre can be processed into insulation batts. Hemp grows at an incredible rate reaching a height of three metres, or more, in 14 weeks. It is sown in the spring and harvested in the autumn. It is left to 'rett' on the ground for a few weeks, to make it easier to separate the fibre. While there is a great deal of controversy about the use of food crops like wheat, rape, palm oil and so on for biofuels, hemp is not grown for fuel or biomass as it is too valuable. Hemp is a useful crop to grow in rotation with food crops such as wheat or potatoes. It cleans up invasive weeds and adds nitrogen to the ground. Hemp does not need any pesticide or herbicide treatment, but some farmers will add fertiliser to the ground. Even if all new houses were built using hemp, this would only require a very small amount of land and hemp can even be grown in poor quality land.

Sheep's wool insulation and other natural fibres:

Sheep's wool is in plentiful supply in the UK and has become popular in housing projects, despite its higher cost. Some sheep's wool insulation is manufactured in the UK though some brands are imported. Natural insulation quilts can also be sourced made from flax, wood fibre, etc.

Hemp and Lime – 'Hemcrete':

Hemcrete is an unusual product in that it is also a building method. Solid walls, floors and roofs can be constructed from a hemp and lime mixture, which is cast like concrete around a timber frame. It provides both insulation and thermal mass and is normally rendered and plastered with a lime plaster. Hemp-Lime construction resolves many of the difficulties experienced in timber frame construction. Conventional timber frame building requires a complicated build up of sheathing and racking boards, breather membranes, insulation and external and internal finishes which have to be anchored to the timber frame. The risk of air leakage is great if membranes are not carefully sealed and fire can spread through cavities if fire barriers have not been properly inserted. The wall remains quite lightweight with little thermal storage capacity. Hemp-Lime on the other hand fully encloses the timber frame providing thermal mass and storage. It protects the timber from damp, rot and fire, reducing the need for chemical preservative and fire retardant treatment. It is much simpler, eliminating the need for breather membranes and can provide the internal finish giving a high level of acoustic absorbency.

Products based on recycled materials:
A number of innovative insulation products are emerging made from recycled materials such as cotton waste and recycled bottles. While these require fire retardant chemical treatment and additives to bind them together they still provide an interesting alternative to completely synthetic insulations. Some of these products emphasis their non-itchy character which is a problem with fibreglass. Some of the unfired earth products incorporate recycled materials such as recovered waste plasterboard.

The application of eco-building?

It might be assumed that the promotion of 'eco-towns' by the UK government will lead to the greater use of natural and eco materials. However there has been growing cynicism about the eco-town concept and some have suggested that it is a ruse to force through more housing development in the countryside (Hunt, 2008; Jenkins, 2008). Those proposals that have been adopted so far are not based on the use of natural or eco materials and rely instead on conventional technology. Bidders planning to use natural materials, like hemp and lime, for eco-town projects have been turned down.

Eco-towns and villages proposed by mainstream developers and the UK government also tend to focus on the use of renewable energy and technological quick fix solutions rather than genuine low impact development. The naivety of this approach was best summed up in an interview in *Prospect Magazine* with Yvette Cooper when she was housing minister in which she called for 'magic wallpaper'.

> 'We need to develop radical environmental technologies that solve problems like that. The way I describe it is that you need "magic wallpaper." The officials think I'm stupid when I say that, but it is what you need – well-insulating, inexpensive stuff that you can put on to solid walls easily and that doesn't add an extra couple of inches to the walls. This is the kind of technology that you need' (Moore, 2007).

While some manufacturers may claim to have developed 'magic wallpaper' with synthetic fossil fuel-based and so-called 'phase-change' materials, these are not low impact solutions to the problem. In reality applying 50mm of hemp and lime plaster to a wall of an old building may have a significant effect in terms of improving thermal performance, but it is unlikely that this is what Cooper had in mind. On the other hand research has begun, led by the Bio-composites Centre at Bangor University, looking at thin natural fibre insulations using natural resins and 'aerogels'. Such materials may be commercially available in a few years.

More radical, smaller and thus less influential groups are trying to pioneer genuine low impact eco developments. Indeed it is possible to identify a growing movement throughout the UK. Many of these groups are taking advantage of innovative social models such as co-housing and community land trusts. Some are supported by the excellent organisation 'Land for People' in an effort to access land at an affordable cost. Such projects are mainly in the countryside but some are also cropping up in inner city areas. Many are driven by people who want to live sustainably on the land by running a smallholding. Such is the demand for this that some rural local authorities like Pembrokeshire have introduced low impact development supplementary planning guidance. However, the strictures of such policies are so demanding that it will not be easy for groups to comply and so far quite a few proposals have been turned down for planning permission. The nature of such groups is that they have few resources and cannot afford a battery of professional consultants and lobbying, unlike mainstream developers. On the other hand, most are committed to using natural and low impact building methods and materials, even sourcing them from their own land if possible (Law, 2005). While such activity may be seen as relatively marginal to housing needs in general, it does reflect a genuine desire among many people to live in a more sustainable way. Such projects can pilot and pioneer natural and low impact building technologies.

Carbon sequestration and role of natural materials in producing zero impact buildings

The introduction of the term zero-carbon houses causes a great deal of confusion when applied to houses that are a long way from being zero-carbon. Most houses, even when extremely well-insulated, normally require a significant amount of power and energy input. Heating for water, background warmth and electricity for appliances, fans and pumps can rarely be avoided except for those who choose to live 'off-the-grid' (Rosen, 2007). The use of renewable energy is far from being zero-carbon as the manufacture and supply of renewable energy requires the expenditure of energy. The construction of buildings also consumes a great deal of energy in the manufacture, transport and installation of materials on site.

There are those who argue that such 'embodied energy' can be disregarded and that reduced energy consumption in use is all that matters, but as resources and energy become more and more costly and precious we need to find ways of reducing carbon emissions in the near future, not simply saying we can reduce them over the 50 or 100 years lifetime of a house.

Inevitably some materials and products that are essential to build (or renovate) a house will use energy, like glass, concrete and steel. Getting the materials to the site will use energy. However, what if sequestering carbon dioxide into the building fabric can offset these? If natural, renewable, crop-based materials are used, then this becomes possible. Timber will lock up a small amount of CO_2 but other crop-based materials such as hemp will dramatically improve on this.

It has been claimed that a Hemcrete wall will lock up over 100kg of CO_2 per square metre (Hemp Embassy, 2008). This claim is based on the amount of carbon dioxide absorbed by the plant material whilst it is growing with the other CO_2 emissions from manufacturing of lime etc. deducted. This can make it possible to save tonnes of CO_2 and store it in houses. Such a form of carbon offsetting is much more credible and cost-effective than paying an organisation to plant trees or buy areas of rain forest.

There are arguments about how to make such a calculation, but current claims may be overly conservative and the sequestration argument is only one of a number of positive features that natural building can offer.

Benefits become even more apparent when Life Cycle Analysis (LCA) is utilised. In future, governments will require building material producers to produce an Environmental Product Declaration (an EPD), which will have to have been carried out by an independent third party. EPDs are based on life cycle methods laid down by the

Plate 9.2: Hemp houses in Haverhill – Suffolk Housing Association, Modece Architects (Photo: Tom Woolley)

International Standards Organisation and will take into account cradle to grave aspects of the material. Natural materials at end of life, when disposed of, can do little if any damage and thus will rate much more highly than synthetic materials.

Natural materials also behave quite differently in buildings to many of the materials that are normally used. Materials like brick, concrete, polystyrene and polyurethane are inert and cannot change their character with changing climate. Throughout the year and each day, temperature, rainfall, solar gain and humidity levels can vary dramatically. Natural materials can respond dynamically to these changes and help to even out temperature and moisture differences. As a result the literature on building and materials science is changing in order to explain the dynamic nature of natural materials (Evrard *et al.*, 2006).

Natural materials have been found to perform better in terms of thermal insulation than might have been expected from normal calculations. Currently building regulations require a 'SAP' assessment to be done into which are inputted thermal conductivity and other values. Thermal conductivity or 'U-values' are based on tests carried out in the laboratory and do not necessarily reflect what happens in a real building in dynamic conditions. Research carried out by the Building Research Establishment (BRE) at a small housing association project in Haverhill, Suffolk, showed that two houses built with Hemp-Lime walls had a better thermal performance than two identical brick and block houses, even though in theory the brick houses should have been better (Suffolk Housing Society, 2008). The reasons for this are complicated and will require further research, however, there are a number of factors that explain this:

1. The ability of the walls to store heat improves internal temperatures.
2. The walls feel warmer to the touch and thus improve thermal comfort for the occupants.
3. The breathability of the walls reduces humidity and this also improves thermal comfort.
4. The ability of the walls to handle moisture improves the efficiency of the fabric.

As a result the occupants feel warmer and tend to use less energy and thus heating bills are lower. Whilst these findings are 'revolutionary', demonstrating the clear advantages over modern methods of construction, it is almost certain that building regulations and calculation methods will not be changed in the foreseeable future. Indeed not all of these materials will behave the same depending on the construction system used. Until standards reflect these new findings, natural materials will be at a market disadvantage.

However, natural materials have other advantages; the significantly better LCA characteristics, improved indoor air quality and so on. Again, current standards such as the 'health and wellbeing requirements' of the Code for Sustainable Homes, do not give sufficient weight to these issues. While there are indoor air quality standards in the building regulations for England and Wales, these standards are so low as to be ineffective. At the moment there is no requirement for houses to be tested to ensure that they are safe and healthy to live in! Indeed, better standards will be opposed by the manufacturers of synthetic materials, as they would be forced to reduce the toxic chemicals used in their manufacture.

Currently, toxic chemicals from flame-retardants, glues and preservatives are emitted into buildings from building materials and finishes like paints, usually referred to as Volatile Organic Compounds (VOCs). Scientific research has shown that these chemicals are highly dangerous to our health and it is relatively easy to detect them in buildings using a VOC meter. Some of these chemicals off-gas at a relatively early stage in the life of a building but others can linger, especially if built into the fabric. In other European countries and the USA, awareness of these issues is much higher than in the UK. Testing is carried out and protection for building workers is much higher. Some medical epidemiological research in the UK has suggested a link between toxic chemicals found in the home and cancer and asthma but again much more work is needed to assess the scale of the problem (Woolley, 2006).

A simple solution to avoid these problems is to use natural materials, particularly those that do not require chemical treatment. Products like 'Hemcrete' which use lime have particular advantages because lime is a natural biocide with high pH content. It not only protects timber from attack, but also prevents mould growth and is anti-static, reducing build up of dirt.

Sourcing natural materials

While many of the products referred to here are manufactured and imported from abroad, a number of key British companies are investing in UK manufacture. Companies like Lime Technology in Abingdon are now producing a specially formulated lime binder called 'Tradical' for Hemp-Lime construction. Hemcore in Essex has built a large new factory for processing hemp. Smaller companies like Natural Fibre Technology in North Wales have pioneered a number of new products. Mainstream distributors like Wolseley and B&Q are following the lead of smaller companies like 'Natural Building Technologies', 'Green Building Store' and 'Ecomerchant' in stocking natural building materials. Large-scale housing and building projects are now incorporating natural and low impact materials throughout the UK. As the supply chain improves then the price and availability of such materials will improve and it will be easier to design and specify with such methods of construction.

Conclusion

In the 1990s it was difficult to document genuinely low impact sustainable and eco construction even though the ideas had been around for some time. However, since the turn of the century there has been a dramatic growth in the development of low impact construction, materials, products and a supply chain. Even though there are many battles to be fought, there is little doubt that much future energy-efficient housing will be built using such materials.

References

BERR (2007) *A Sustainable Construction Strategy*, Department for Business, Enterprise and Regulatory Reform. Available from: http://www.berr.gov.uk/consultations/page40642.html

British Gas (2006) *Tudor architecture surprisingly green compared to modern homes*. Available from: http://www.britishgasnews.co.uk/index.asp?PageID=19&Year=2006&NewsID=698

Evrard A., De Herde and Minet, J. (2006) 'Dynamical Interactions between heat and mass flows in Lime-Hemp Concrete', in P. Fazio, H. Ge, J. Rao and G. Desmaris (Eds), *Research in Building Physics and Building Engineering*, London: Taylor and Francis

Hemp Embassy (2008) *Building with Tradical Hemcrete*. Available from: www.hempembassy.net/hempe/resources/BuildingwithTradicalHemcrete.pdf

Hunt, T. (2008) 'The un-eco eco-towns – The government has entered into a pact with developers – and our countryside is suffering', *The Guardian*, 23 February

Jenkins, S. (2008) 'Ecotowns are the greatest try-on in the history of property speculation', *The Guardian*, 4 April

King, D. (2007) 'At last, I'm hopeful about climate change', *The Independent*, 1 January

Kochan, B. (2008) 'The Ins and Outs of Insulation', *Sustainable Building*, February Issue 14 pp.8-9

Land for People. Available from: http://www.landforpeople.co.uk/

Law, B. (2005) *The Woodland House*, Permanent Publications

Moore, R. (2007) Available from: www.prospect-magazine.co.uk/article_details.php?id=9922

NNFCC (2008) Available from: http://www.nnfcc.co.uk

Rosen, N. (2007) *Off-Grid – How To Live Off-Grid Journeys Outside The System*, Transworld Publishers Ltd

Ryan, C. (forthcoming) *Traditional Building Construction and its application to the construction of sustainable new build*, London: Taylor and Francis

Suffolk Housing Society (2008) Available from: http://www.suffolkhousing.org/

Woolley, T. (2006) *Natural Building*, Crowood Press

Yu, C. and Crump D. (2007) (BRE) *Indoor Air Quality Criteria for Homes for Assessing health and Well-being*, proceedings of the 8th International conference of Eco Materials ICEM 8, Brunel University UK, 9-11 July 2007

CHAPTER 10:
Climate change and the existing housing stock

Gavin Killip

Introduction

Energy conservation is a common theme in existing policy on energy in housing and in debates about climate change. The policy approach to date has been to subsume climate change targets under the pre-existing energy-efficiency programmes, making the assumption that progress on tackling fuel poverty and improving health and social welfare will be enough to combat climate change as well.

In fact, energy-efficiency policies are inadequate for the task in hand (and display a lack of clarity about targets and definitions of savings). The policies motivated by improving social welfare are not enough to help the UK meet its CO_2 reduction targets, but a more radical agenda for change would be. Both the climate change targets and the social housing objectives can be met, but only if we raise the ambition of the policy objectives, clarify the real potential, and look beyond grants for one-off measures such as cavity wall insulation, which meet a narrow definition of cost-effectiveness. A new system and infrastructure for whole-home carbon audit and refurbishment is needed to make UK homes fit for the twenty-first century. Lessons can be learned from abroad, particularly Germany, although even the more ambitious programme there seems to be off target after its first decade in operation.

A strategic investment of several billions of pounds per year is needed to make the difference, although not all of the money needs to come from the public purse. Financing change is one challenge. Others include providing an adequate regulatory framework, reinforcing the provision of information on energy performance in the housing market; finding ways to improve and maintain quality in the construction sector through training for new skills and better enforcement of standards. Concerns about architectural heritage need to be addressed, as do environmental impacts beyond energy. Ultimately, there is only a point in achieving a low-carbon housing stock if it also represents a built environment that supports strong communities, healthy local economies and a good quality of life for UK citizens.

Current UK approach to energy efficiency

Confusion about 'savings'

Confusion about progress in reducing energy consumption and CO_2 emissions arises from the use of the word 'savings' in official documents and government targets to mean two different things. In relation to new-build housing, for instance, the government's Energy Efficiency Action Plan 2007 claims that revisions to the building regulations in England and Wales in 2005/2006 will achieve savings of 2.6 million tonnes of carbon (MtC) by 2020 (Department for Environment, Food and Rural Affairs, 2007). This use of 'saving' means that emissions will be lower than they would otherwise have been without the revision to the regulations. It describes a reduced rate of growth in emissions, but growth nonetheless. This meaning can be termed 'relative savings'. Growth in emissions from new housing is linked to a complex set of questions about population and demographic change, and to the state of the housing market, housing shortage and a host of social and economic consequences, which new development can improve. There are good reasons, therefore, for wishing to build more homes but, in purely environmental terms, it is misleading to associate a growth in the number of buildings with a 'saving' in CO_2 emissions. The word 'saving' can also mean actual reductions from one year to the next, not relative to how things might have been or measured per building or per unit of floor area, but in absolute terms. Absolute savings are achievable and can be large, particularly where the system in question is highly wasteful to start with.

By treating these relative and absolute savings equally, the Energy Efficiency Action Plan is misleading about what can be achieved with current policies. In fact, 37 per cent of the quantified 'savings' in this document come from policies for new housing, and should properly be described as sources of growth in emissions (Killip, forthcoming).

In debates about housing and the environment, the scope for making absolute savings in energy consumption lie with the existing housing stock, while reductions in CO_2 emissions can derive from both energy savings and new forms of energy supply, such as renewable energy technology.

Across the UK housing stock space heating dominates other end-uses (59 per cent of the total), with hot water accounting for 24 per cent and cooking just five per cent. Lights and appliances made up 12 per cent of the total in 2002, but this sub-sector of demand has been growing rapidly (Department of Trade and Industry, 2002). Significant reductions can be made to reduce heat loss with insulation (ground floor, external walls, roof) and controlling ventilation (e.g. blocking up chimneys, draught-proofing windows and doors). Replacing heating appliances with more efficient models can make big reductions for heating space and hot water (as most heating systems do both). When it comes to lights and appliances, technological improvements have been achieved at the European level, with minimum standards for efficiency applied to manufacturers of light bulbs, fridges and other domestic electrical equipment. In addition to these energy-efficiency measures, there is a range of low- and zero-carbon technologies that can be integrated into homes.

Household energy consumption and energy-efficiency policies

Between 1970 and 2000, energy consumption in the residential sector rose by an average one per cent per year, as the aggregate result of competing forces, some tending to increase energy consumption and some tending to reduce it (Table 10.1).

Table 10.1: Forces for growth and reduction in UK household energy consumption, 1970-2000

Causes of growth	Causes of reduction
Higher internal temperatures	Increased boiler efficiencies
Increased population	Better controls[1]
Smaller households (fewer people per dwelling)	Insulation measures – lofts, cavity walls, hot water tanks[2]
More electrical appliances per household	
Changing washing habits, more hot water	

Sources: Department of Trade and Industry, 2002; Shorrock and Utley, 2003.

Energy-efficiency programmes over this period kept the growth in energy demand from being as high as it might have otherwise been, but the overall trend has been steadily upwards (Figure 10.1).

Figure 10.1: Energy consumption from the UK residential sector, 1970-2000, showing savings from energy-efficiency policies (hatched)

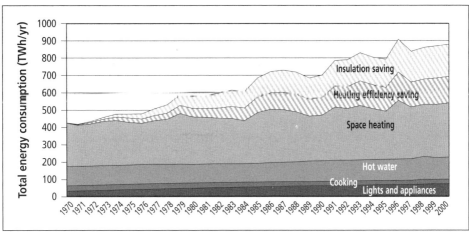

NB: Irregular shape of the space heating consumption reflects variations in winter temperatures.
Source: Department of Trade and Industry, 2002.

---

1   Savings from controls assume good user understanding and 'correct' operation, but this is often not achieved in practice. Better feedback is needed from actual users to the designers of control technology, as well as better training and instructions for users.
2   Quality of installation needs to be high if savings are to be achieved in practice.

The history of energy-efficiency policy in UK housing has been one of modest improvements, motivated primarily by the concern to tackle the health impacts of cold, inefficient homes on vulnerable people and to do so in a cost-effective way. Thus, the measures promoted through advice services and supported by grants are typically limited to those with a simple payback of less than seven years: cavity-wall insulation, loft insulation, efficient central heating systems and controls, hot water tank jackets, draught-proofing, low-energy light bulbs (compact fluorescent lamps or CFLs). The principal programme is funded by an obligation on energy suppliers to invest in energy efficiency, known as the Energy Efficiency Commitment (EEC) (renamed Carbon Emissions Reduction Target (CERT), for the period 2008-2011). In the three years 2002-2005, EEC '...*stimulated about £600m worth of investment in energy efficiency*' (Department for Environment, Food and Rural Affairs, 2007).

A persistent problem is fuel poverty,[3] and investment in energy-efficiency improvements has been targeted at these most vulnerable households through the Decent Homes Standard, Warm Front in England and similar programmes across the UK. Following recent energy price rises, fuel poverty affects approximately four million UK households (Boardman, 2007) or roughly 15 per cent of the 25.7 million total. However, 50 per cent of energy-efficiency savings under EEC had to come from a group of low-income consumers, while current proposals for CERT (a substantially bigger programme than EEC2) are for 40 per cent of carbon savings to come from a priority group of low-income and elderly consumers. Energy supply companies point out that the objective of tackling climate change would be better served by targeting higher energy consumers, and that the continuing focus on fuel poverty under EEC and now CERT is evidence that the main focus of this policy instrument is alleviating fuel poverty, not reducing environmental impacts (House of Commons Communities and Local Government Committee, 2008).

Investment to alleviate fuel poverty makes perfect sense from a social welfare perspective, but the targeting of relatively limited funds on fuel poverty will do very little towards achieving deep CO_2 emissions reduction targets. Instead, what is required is a radical shift in policy towards a programme of housing refurbishment which can both eradicate fuel poverty and make a significant impact on the sector's total CO_2 emissions. This will necessitate a new perspective, with policies and incentives aimed at improving the housing stock for the more affluent, 'able-to-pay' householders as well as for the vulnerable groups.

In addition to expanding the target market for this work, the list of measures needs to be expanded beyond what is currently supported, to include all technically feasible refurbishment opportunities (Table 10.2).

---

3   A household is defined as being in fuel poverty if total expenditure on energy exceeds 10 per cent of disposable income. For a detailed discussion of fuel poverty see Chapter 6.

Table 10.2: Energy-efficiency refurbishment measures, showing which are supported by policy and which are not

Measures supported under EEC/CERT	Other technically feasible refurbishment measures
Cavity-wall insulation	Solid wall insulation
Loft insulation	Ground floor insulation
Draught-proofing	High-performance glazing
Hot water tank insulation	Reducing air infiltration (e.g. blocking up redundant chimneys, flues)
Efficient heating boiler[4]	Passive solar design features (where site conditions allow)
Heating controls	

Source: Killip, forthcoming.

Low- and zero-carbon energy technologies in the built environment

There are a number of energy technologies that can be incorporated into the built environment, which either use fossil fuel in a particularly efficient way or else capture renewable energy from the immediate environment. Some provide heat, some provide electricity, while the Combined Heat and Power (CHP) technologies provide both (Table 10.3). As a group, these are known as low- and zero-carbon technologies (LZCs) (the terms 'micro-generation' and 'micro-generators' are also sometimes used, although heat pumps are not strictly speaking a generation technology).

Table 10.3: Classification of low- and zero-carbon technologies by type of energy output and environmental impact compared with conventional systems

	Heat only	Electricity only	Heat and electricity
Low-carbon technologies	Heat pumps		Combined Heat and Power (CHP)
Zero-carbon technologies	Solar water heating Biomass stoves	Photovoltaics (PV) Micro-wind	Biomass CHP

LZCs have been promoted through a series of different grant schemes. The Low Carbon Buildings Programme was launched in 2006 and is due to run until spring 2008. By November 2007 it had allocated some £7.1 million (Allen, Hammond *et al.*, 2007).[5] Even taking into account the number of installations that existed before these grants became available, the total number is very small in relation to the size

---

4 Condensing boilers were supported under early years of EEC, but have been ineligible for support since they became mandatory under minimum standards introduced in 2005.
5 Scotland has had the Community and Household Renewables Initiative (SCHRI); the Renewable Energy Fund has existed in Northern Ireland.

of the stock (Table 10.4). In relation to the LCBP, Allen *et al.* conclude that, '*...it is unlikely that the amount of funding available will stimulate the market sufficiently to lower the capital costs of micro-generators in the near future*' (Allen, Hammond et al., 2007).

Table 10.4: UK domestic stock of LZC technologies in 2005 and projects funded under the Low Carbon Buildings Programme (LCBP)

LZC technology	Total number (2005)	No. projects funded by LCBP
Solar thermal	78,470	2,122
Solar PV	1,301	510
Micro-wind	650	1,493
Ground source heat pumps	546	274
Micro-CHP	990	–
Micro-hydro	90	4
Biomass room heaters	–	16
Wood fuelled boilers	–	116
Total	82,047	4,535

Source: Adapted from Allen, Hammond et al., 2007.

The Home Truths report commissioned by Friends of the Earth and the Co-operative Bank estimates that just over 100,000 homes were served by low- and zero-carbon technologies in 2007, including some 25,000 homes on community Combined Heat and Power (CHP) schemes (Boardman, 2007). With 25.7 million homes in 2007, this amounts to roughly one LZC for every 300 homes. In the *40% House* study, a 60 per cent reduction in CO_2 emissions by 2050 was achieved by energy demand reduction measures beyond what is currently cost-effective and an average of between one and two LZCs per home (Boardman, Darby et al., 2005). In other words, current installed capacity is perhaps 0.4 per cent of what is required to meet the CO_2 reduction target. Incremental changes to current policies may increase uptake to a certain extent, but they are not sufficient to meet the scale of the new challenge.

Comparison with Germany

The UK experience contrasts with that of Germany, where the sums of money are considerably higher and the programmes are differently designed. The Kreditanstallt für Wiederaufbau (KfW)[6] provides low interest loans towards major refurbishment of pre-1984 dwellings, aiming to bring them up to the standard of current new build. The average loan amount in the period 1996-2004 was over €20,000 per dwelling (over £14,000) for the rehabilitation programme (for an integrated package of building improvement) and €8,300 per dwelling (over £5,700) for the CO_2 reduction programme (for simpler add-on measures) (Chatterjee, 2007).

---

6   KfW is the German government-funded development bank, created after the second world war to support reconstruction projects in East Germany.

The target reduction by 40 kg CO_2/m^2 under the German rehabilitation programme is more ambitious than anything attempted to date in the UK, representing a cut of approximately 50 per cent in CO_2 emissions. The pace of change is also ambitious, although the published aim to refurbish five per cent of the housing stock per year is far from being achieved in practice.

The German approach costs about £750 million per year (over three times the sum available in the UK) and has covered 881,000 buildings in nine years. With some 17.3 million residential buildings, this equates to 0.6 per cent of the stock refurbished per year as a nine-year average. To put this into perspective, it would take 177 years to treat the entire German housing stock at this rate of improvement.

The German record on installing micro-generation technologies is more positive, as it is one of several dozen countries to offer a 'feed-in tariff' for building-integrated renewable energy systems. Unlike UK support programmes, which have all offered a grant towards the capital cost of installation, a feed-in tariff guarantees a premium price for every unit of renewable electricity generated over a guaranteed period (i.e. 20 years). In Germany, this has been instrumental in incentivising the installation of some 300,000 solar photovoltaic (PV) systems and a thriving new industry with 150,000 jobs and an annual turnover of €12 billion (Department for Business Enterprise and Regulatory Reform, 2006).

The comparison with Germany offers insights for developing a policy for low-carbon housing in the UK. Firstly, there is the existence of the German reconstruction bank, the KfW, and the willingness of the German government to make low-interest finance available by increasing public borrowing. Secondly, a much larger proportion of German housing is in apartments than is the case in the UK, which means that large-scale interventions on entire apartment blocks can make a bigger impact on the condition of the stock. Thirdly, the packages of measures supported through the German loan scheme are designed to achieve deep cuts in one go; a much more ambitious and integrated set of measures than the ones that the UK promotes within its narrow definition of cost-effectiveness. Finally, the fact that the more ambitious German model is off target by a factor of about ten underlines how difficult this transition is to manage. It may be that the missing element in the German scheme is a minimum standard. There is a range of incentives and sources of information but no penalty for non-participation in the scheme. On micro-generation, the UK government acknowledges the success of feed-in tariff arrangements in Germany and elsewhere, but this approach is seen as 'too interventionist' in the UK's liberalised energy market.

Energy Performance Certificates (EPCs)

The EU's Energy Performance of Buildings Directive has been instrumental in making available and visible the basic information on energy performance of buildings. In the UK residential market, this involves the requirement to provide an Energy Performance Certificate (EPC) when a property is sold (rentals are due to be included from October 2008). Based on the familiar A-G rating displayed on domestic appliances, the EPC gives an indication of the current condition of the property along with its potential for improvement using cost-effective measures (Figure 10.2). The report accompanying the EPC also lists a range of additional measures that could be adopted, including a small number of LZCs.

Figure 10.2: Example of an Energy Performance Certificate for England & Wales

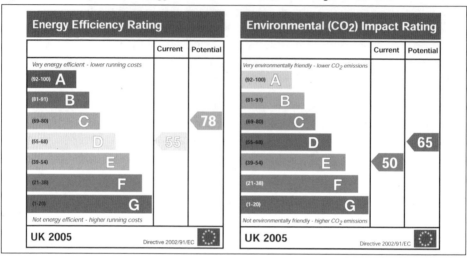

The EPC has the potential to be a powerful tool but that potential will only be realised fully if other policies are brought into play (for a more detailed discussion on the effectiveness of EPCs in the UK context see Chapter 16). So, for example, the visibility and impact of the certificate would be greatly enhanced if it had to be displayed on all marketing materials (i.e. in printed media, on property websites, in estate agents' windows) rather than being provided in an information pack, which has to be requested from the seller/agent. The success of the policies on electrical appliances is often attributed to the introduction of an energy label, perhaps because this was the most obvious, visible change that occurred. In fact, the success was due to a set of policies, of which the energy label was only one. Incentives and minimum standards also played their part (Fawcett, Lane et al., 2000).

Transformation of the UK housing stock

Targets and scenarios

The UK's 2003 Energy White Paper set out four key objectives for energy policy in the twenty-first century (Department of Trade and Industry, 2003), namely:

- achieve a 60 per cent reduction in CO_2 emissions by 2050 on a 1990 baseline;
- increase security of energy supply;
- boost UK competitiveness; and
- eradicate fuel poverty.

More recent climate science suggests that the reduction target may need to be as high as 80-90 per cent in industrialised nations in order to achieve stabilisation of atmospheric CO_2 at safer levels in terms of the risk of catastrophic climate change (Tirpak, Ashton *et al.*, 2005). The 60 per cent reduction target has provided the impetus for a number of scenarios, using computer models to map the possible future impacts of housing based on a series of assumptions about the key variables and parameters. While these scenarios are not intended as predictions of the future, they do help clarify both the scale of the challenge and the inter-dependency of the variables involved. Different 'solutions' can be found to the problem of achieving a 60 per cent reduction in CO_2 emissions, as this brief summary illustrates.

A team from Leeds Metropolitan University's Centre for the Built Environment developed several scenarios, including an 'integrated' scenario, which combined demand reduction measures with a much reduced carbon intensity of grid electricity to achieve well over the 60 per cent reduction target (Johnston, 2003; Johnston, Lowe *et al.*, 2005). Notable among the measures put forward here was the assumption that all older, solid-walled homes could have their walls fitted with 150mm insulation. A team led by the Environmental Change Institute at Oxford University developed only one scenario (achieving 60 per cent), which included the uptake of refurbishment measures beyond what is currently cost-effective, including insulation of one million of the seven million existing solid-walled homes and a significant take up of low- and zero-carbon technologies, as well as a four-fold increase in the current rate of demolition (Boardman, Darby *et al.*, 2005), which can be beneficial in life-cycle carbon terms if the quality of what gets built to replace the older homes is high enough (Royal Commission on Environmental Pollution, 2007).

Figure 10.3: CO_2 emissions from refurbished and new-build housing in a scenario achieving 75% CO_2 reductions by 2050

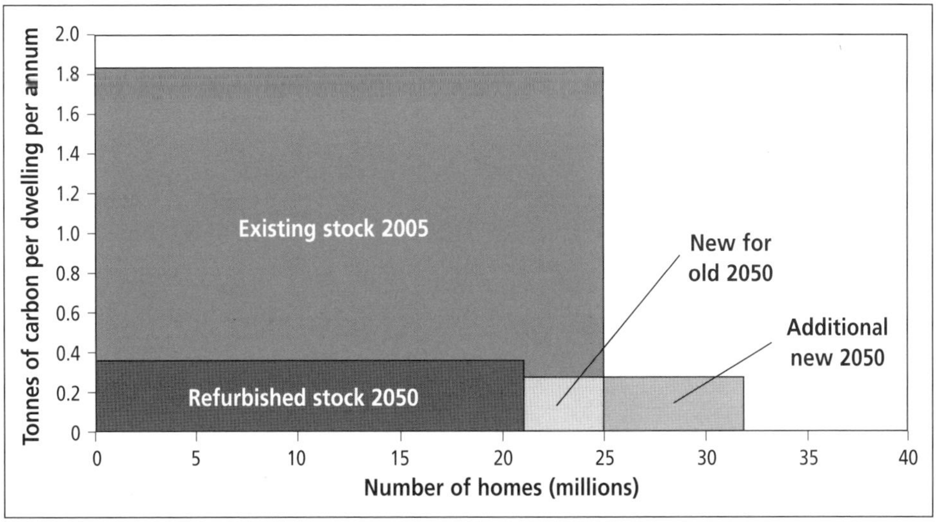

Source: Royal Commission on Environmental Pollution, 2007.

What these scenarios have in common is a basic assumption that radical change is needed in order to reach or exceed the 60 per cent reduction target by 2050. The shift of perspective is profound; those homes that are conventionally described as 'hard to treat' among housing and energy professionals actually have a greater potential for improvement in purely technical terms (Lowe, 2007). The challenge is to conceive of a housing and energy policy framework, in which a significant fraction of this potential could be realised.

From niche to mainstream

A small number of advanced refurbishment projects already exist in the UK (see, for example Climate Outreach Information Network, 2007, the Environmental Change Institute, and the Sustainable Energy Academy). They demonstrate clearly that deep cuts in environmental impacts are possible (including CO_2 emissions, mains water consumption, waste and toxic pollution), whilst simultaneously improving comfort and quality of life. In order to achieve improvements across the housing stock, this kind of refurbishment needs to move out of its current niche market and into the mainstream. This means creating a framework in which all the various agents involved in housing play their part – from private homeowners to public and private landlords; from builders, plumbers and other tradespeople involved in refurbishment to new businesses, such as installers of solar panels; from estate agents and surveyors to mortgage companies.

The scale of the challenge and the need for a truly 'joined-up' strategy is highlighted in the *Home Truths* report, which explores the policy implications of an 80 per cent reduction target (Boardman, 2007). This argues for tough standards and legally-binding targets devolved down from national to local government. Energy performance certificates are seen as more than just a source of information: the A-G ratings are proposed as the base for a system of minimum standards, getting tougher over time. If this scheme were to be implemented, any purchaser of a wasteful, inefficient property would be required to upgrade it before it could be re-sold. With rental properties, the onus to upgrade would fall on the landlord: properties failing to meet the minimum standard could not be let. Property transactions are seen as a key intervention point for several reasons: firstly, this is a process which already requires legal and financial oversight, from the transfer of property deeds, to mortgage lending, to registration for stamp duty. Secondly, with 1.5 million property transactions per year in the UK, there is sufficient activity to make more than a small impact on the overall problem, especially if each intervention is built around a suitably ambitious target. The idea of regulating property transactions to make low-carbon refurbishment mainstream has its own problems, notably a distinct lack of political support. Also, the average figure for the 'churn' rate of housing hides the fact that a significant minority of homes are occupied for a very long time before being sold on, and these homes would pass through the net of a system triggered only by transactions (Boardman, 2007). Another route, and one that might be easier to introduce first in terms of political acceptability, would be to extend the coverage of building regulations to cover so-called 'consequential works' when planning consent is sought for major refurbishments or extensions. This would mean a condition of planning approval for the new extension would be that a minimum standard be met in the pre-existing building. For example, it would be impossible to get approval for a loft conversion if cavity walls in the original building were uninsulated, among other energy conservation measures.

The key shift in emphasis is moving away from piecemeal activity applied one measure at a time, and towards a whole-home refurbishment process with a quantified outcome that represents a reduction in CO_2 emissions, and which is consistent with national targets. Following the German model, this could be based on a standard package of measures for common dwelling types (e.g. solid-walled pre-1919 terrace houses, post-1930 cavity-wall semi-detached houses). For less common dwelling types and cases where the standard package of measures was deemed unsuitable (for whatever reason), a bespoke standard could be worked up by an expert setting out design options and calculating potential impact reductions. A commitment to disseminating results, be they good or bad, should be part of an iterative process of learning, experimenting and further innovation.

Application of such a refurbishment standard would need to be supported by other policies, including innovations in billing and metering to improve information on energy consumption, and financial incentives to make the very considerable costs more bearable. This might include rebates on tax (possibly council tax or stamp duty) or additional lending for refurbishment through 'green' mortgages, which manage the risk to lenders on the principle that money saved through reduced energy costs can be used to pay slightly higher monthly mortgage payments. Refurbishment can also be part of wider, district-based solutions, such as district heating and community combined heat and power (CHP) These communal solutions can achieve economies of scale compared to single-dwelling systems and they are common in several European countries, particularly in Scandinavia.

Costs

The Sustainable Development Commission has estimated that advanced refurbishment of this kind would cost in the region of £25,000-£30,000 per dwelling, while the Environmental Change Institute give a range of £20,000-£60,000, depending on the choice of LZC technologies installed (House of Commons Communities and Local Government Committee, 2008). Costs of individual technologies have the potential to reduce very significantly when starting from a low level of adoption, which is the case especially for many of the LZC technologies, such as solar photovoltaics, which are currently among the most expensive options

(Hinnells, 2005; Hinnells, 2006). Where major refurbishment works are being carried out anyway, the marginal cost of energy-related works can be drastically reduced to, perhaps, 10-20 per cent of the total project cost. In fact, the marginal cost of advanced, low-carbon refurbishment is broadly on a par with the total VAT element of refurbishment, which is at 17.5 per cent on most items. The Cut the VAT Coalition comprises a wide range of organisations supporting VAT reform to make refurbishment costs more attractive, especially in comparison with new construction, which is zero-rated for VAT.

The cost of low-carbon refurbishment for the entire housing stock is likely to be several billion pounds per year: the *Home Truths* report puts it at £13 billion per year (Boardman, 2007), although it is not clear how much of that cost could be reduced through taking advantage of marginal cost effects, nor who is likely to pay. In total, some £23 billion is spent each year on repair, maintenance and improvement in the UK housing stock (Department of Trade and Industry, 2006), a figure well in excess of what is being discussed for low-carbon works. In more affluent areas, certainly, there is plenty of anecdotal evidence of dwellings being bought and renovated, with substantial sums being spent on extensions, conversions and new kitchens and bathrooms. Each of these renovations which fails to take on board the low-carbon agenda – and that is pretty much all of them – can be viewed as a wasted opportunity. This is where innovations – not just new technologies, but also new techniques using well-proven technologies – could begin to make an impact.

Embodied energy

If the aim is to reduce overall impacts, it makes sense to widen the scope to include the so-called embodied energy in all the insulation, solar panels etc. which are being championed by advocates of low-carbon housing. If it takes more fossil energy to make the product than is saved during that product's lifetime, the overall impact is clearly negative. These life cycle assessments involve some degree of uncertainty and crucially hinge on the assumptions made. So, for instance, it is important to get an accurate forecast of annual output from a solar panel and a good idea of its lifetime before it is possible to compare the energy likely to be generated by the panel with the energy cost of its manufacture, transport to market and subsequent decommissioning. Estimates for the embodied energy of low-carbon refurbishment do vary but a review of available figures suggests that the total is about 15,000 kWh (Environmental Change Institute, 2007). The average figure for annual residential consumption is about 22,000 kWh, and advanced refurbishment could cut that by 50 per cent or more (Boardman, Darby *et al.*, 2005; Johnston, Lowe *et al.*, 2005). If savings of 11,000 kWh per year are assumed, then the embodied energy of the refurbishment is 'paid for' in a little over a year. The vast majority of refurbishments will not be ripped out that quickly so, in purely energy terms, the work has an overall positive impact.

Education and human behaviour

The impact of resident behaviour on energy consumption is very large.but, because it is seen as uncontrollable by building scientists, it tends to be ignored or, at best, tackled as an afterthought. A survey of 29 new zero-carbon homes in California, all built as part of a single development to very similar designs, showed huge variation in monitored energy consumption. The lowest consuming home actually exported renewable energy, while the highest consuming home used double the average for the state of California (Janda, 2007; Keesee, 2005).

Education and awareness-raising activities typically include provision of written information (e.g. leaflets), websites and other media, and public engagement through events and advice services. School students are often seen as a key audience.

Studies have shown that there are other modes of communication that can help reduce energy consumption. While energy meters remain tucked away in locked boxes or under the stairs, no one pays them much attention, but trials with visible, user-friendly displays in areas which are used regularly (e.g. the kitchen) can be effective in changing consumption patters. Direct displays, new kinds of energy meters and better billing can lead to a reduction of 5-15 per cent in consumption (Darby, 2006). The UK government aims to roll out better display and metering technology across the residential sector, but plans have been dogged by the question of who is to pay. Ownership of building-integrated renewable energy technologies can also have a beneficial effect on the awareness and behaviour of residents. One study found that having photovoltaic solar panels on the roof not only produced green electricity for the householder, but also led to a six per cent reduction in electricity consumption. In other words, there seems to be a causal link between the presence of the technology and a measurable change in behaviour (Keirstead, 2007).

Quality, performance, training and skills

Real-life energy performance of buildings rarely matches up to the design. One key reason for this is the serious 'blind spots' in design and construction practice which lead to much higher levels of heat loss than expected (for example Lowe and Bell, 2000; Olivier, 2001; Roberts, et al., 2005, see also Chapter 8). A second is the lack of systematic feedback during the various processes – from design to construction to final occupation and operation (Bordass, 2005). With almost 50 per cent of the UK's CO_2 emissions coming from buildings (27 per cent from housing), the need to monitor and improve performance of buildings is one of the key challenges of climate change policy, and one that goes largely forgotten or ignored.

With micro-generation technology, there are comparable 'devil in the detail' issues which need to be more widely understood and acted upon. For instance, electrical systems (e.g. solar photovoltaics, wind turbines) may 'trip out' to protect equipment from surges in output and need to be re-set if any renewable energy is to be collected. Providing a visible signal that the system needs re-setting is crucial if months or years of non-generation are to be avoided. Similarly, solar water heaters may provide little or no useful energy if a conventional heating system has been used to heat the tank up just before sunrise. These potential pitfalls for new technology may seem daunting, but it is important to bear in mind that similar issues arise for any new technology. The motor cars and gas boilers which we today see as reliable and efficient, with a service industry to carry out repairs and maintenance, are the products of years of development. The creation of new markets will involve not only sales of equipment, but also new categories of support infrastructure, creating new jobs requiring new skills.

Learning is clearly needed, but so is a complementary system of enforcement and sanctions. It may be that builders do not appreciate fully the importance of snugly fitting insulation, and that training could address that gap in knowledge. However, it is also true that builders systematically choose to take more care over the finished appearance of a wall than on getting the insulation inside it to fit properly – precisely because they know which aspects of their work will be inspected most closely, and which will be ignored. Corner-cutting may be tackled more effectively through the threat of sanctions than simply through training.

In the area of LZCs, the construction industry faces an issue of cross-skilling or multi-skilling. Where traditional building trades have been quite specialised, and a system of sub-contracting has been used to keep each trade separate, the prospect of fitting new technologies such as solar panels cuts across these traditional boundaries. A solar water heating fitter needs to know aspects of four traditional trades (roofing, plumbing, electrics, general building) as well as the specialist skills of siting, sizing and fitting the solar collector.

Conservation of architectural heritage

The tensions between a desire to preserve old buildings and conserve energy in order to reduce CO_2 emissions have been brought to the fore with the advent of Energy Performance Certificates in Home Information Packs. Many energy conservation measures (e.g. wall insulation, replacement windows) are seen by the heritage movement as detrimental to the historic character of well-loved older buildings, and therefore to be resisted. English Heritage, in its interim guidance to Domestic Energy Assessors (DEAs – the people who survey homes and produce energy performance certificates) argues that the starting point for assessors should be an assumption that no fabric measures can be incorporated in what they call historic and traditional buildings – about 25 per cent of the stock (English Heritage, 2007). Any fabric measures that do get recommended, it is argued, should be justified on a case-by-case basis. Similarly, roof-mounted solar panels and other visible renewable energy technologies are subject to quite tight constraints in conservation areas.

Plate 10.1: This normal-looking terraced house has been renovated to achieve a 60 per cent reduction in CO_2 emissions without sacrificing comfort.

See www.ecovation.org.uk for details.

(Photo: Gavin Killip)

The effect of removing 25 per cent of the stock from a programme of fabric improvements would be to make deep cuts in CO_2 emissions that much more difficult. Many of the period details that conservationists seek to preserve, such as cornicing, can be re-created on top of internal wall insulation. The approach of the Nottingham Ecohouse (a refurbished Victorian semi-detached property) uses internal wall insulation (and re-created cornicing) on the front façade of the house and external insulation on the side wall and rear extension. Thus, the Victorian brickwork and appearance of the streetscape remain unchanged, but the walls away from the façade do change in appearance, from brickwork to rendered insulation. A similar approach can be taken with windows, with double-glazed panes in sash frames at the front (preserving the appearance) and higher-performance (non-sash) replacement units at the back and sides (Poyzer and Schalom, 2003).

A public debate, informed by research and practical demonstration, is needed to resolve the current conflict between heritage conservation on the one hand and environmental protection on the other. The Nottingham Ecohouse is one example of an advanced refurbishment that has been carried out with considerable regard for aesthetics and preserving architectural heritage, but heritage groups seem to view the low-carbon agenda as a threat to architectural conservation. This debate is only in its infancy but it has already got bogged down in an unwillingness to accept change among conservationists. The possibility that past practices and attitudes may not be suitable in future seems to meet with a strong negative reaction despite the fact that it is the flexibility and adaptability of older homes – which their advocates often cite as one of the reasons why they are so well-liked – that actually makes them suitable for extensive eco-renovation (Lowe, 2007). It seems ironic, therefore, that those same advocates should be so resistant to change.

Wider sustainability issues

The climate change agenda is a political priority but there are clearly other important issues to be considered when creating sustainable urban environments. Other environmental impacts of buildings besides energy and carbon need to be taken into account (water, waste, pollution, biodiversity), and the environmental agenda needs to be addressed within a context of sustainability. Good homes need to be part of 'good' neighbourhoods, with strong communities and healthy economies. It is beyond the scope of this chapter to discuss urban design issues in depth, but one of the arguments for bringing older housing up to a low-carbon standard is that older neighbourhoods often have many of the sustainable design features that are so lacking in modern 'close' developments: traditional streets allow for public and private open space, a markedly higher density of housing than in most modern developments, leading directly to more vibrant local high streets and viable public transport systems, as well as shorter distances between destinations, which make walking and cycling more practical for the majority. The provision of local shops, schools, hospitals and leisure facilities is generally better as well – all leading to a greater sense of place and community focus.

Roles and relationships in refurbishment

Housing professionals face a new set of challenges in managing their stocks of buildings if this low-carbon agenda is to become mainstream. Many of the technologies and ways of doing things described here may be new and unfamiliar, leading to uncertainty and fear that targets on paper are not achievable in practice. It is easy to be seduced by the technology of sustainable housing refurbishment, and there is no doubt that housing professionals will have to become more familiar with new products and techniques, many of which have undeniable 'wow factor'. However, it is worth preserving some detachment from the technical detail, as this should rightly be the preserve of technical consultants. The housing professional, as the client for refurbishment work in the public sector, has an important role to play in two areas: firstly, to ensure that the relationships in the design and construction team are working well, with contractual relations that encourage collaborative problem-solving, through partnering and a culture of open communication between them at all stages of a project; and secondly, as custodians of the overall project objectives. If refurbishment work is to achieve enough, there is an important role for the client to set tough but achievable targets for environmental performance, and to champion those standards consistently throughout the project. If the designers' preferred technology (or group of technologies) in the design is shown to be inadequate to the task, the role of the client should be to guide the design team back towards the drawing board, not to get involved in narrow debates about the detail of one technology or another.

Towards a low-carbon refurbishment strategy for the UK

There are probably a few hundred UK homes that have been refurbished to a high energy standard, of which a few dozen are currently publicised (Climate Outreach Information Network, 2007; Poyzer and Schalom, 2003; Royal Borough of Kensington & Chelsea, 2007; Sustainable Energy Academy, 2007). These advanced refurbishments have mostly been carried out by private individuals as an expression of their strong environmental values and a desire to live in warmth and comfort without wasting resources.

While technical innovations may play a role in the future, the barriers to wider uptake of such eco-refurbishment ideas are not due to a lack of technology: most of the energy-saving technologies that work as retro-fit options have been in existence for decades. The real barriers are to do with information, cost, disruption, the lack of capacity to deliver among the construction industry and, arguably, culture: too few UK citizens have experience of energy-efficient buildings, so the current level of wasted energy and poor indoor comfort is accepted as normal.

In order to make eco-refurbishment mainstream, concerted efforts are required on several fronts, including the following:

- do demonstration projects – prove what is possible and use real homes to communicate at all levels: for media and PR work, educational visits and as a training resource for businesses involved in refurbishment;
- monitor and evaluate – measure the environmental impacts (ideally 'before and after') and build up a body of data to prove what works and to inform future research needs;
- develop a refurbishment standard which is ambitious, achievable and grounded in monitored evidence;
- finance innovation in the social housing sector (housing associations are repeat clients for building work and have the experience and capacity to deliver);
- initiate training for skills in collaboration with manufacturers and other industry stakeholders, based on ongoing evaluation of which materials, designs and techniques work well in different situations;
- increase awareness of energy issues in housing by widespread use of the Energy Performance Certificate (EPC) and better metering and billing information for householders;
- set minimum standards for selling/renting property, e.g. make it impossible to sell or let a property rated F or G from 2013; and
- set a clear timetable for future minimum standards so that, say, it is illegal to sell or let a property rated lower than C by 2020.

Some of this will require policy from national government, but there is much that can be done at the local and regional level – and done much faster than it takes for national policy to be implemented. Local planning policies have been used in innovative ways to require renewable energy technologies in new developments (an approach pioneered by London Borough of Merton in 2004, sanctioned in Planning Policy Statement 22, and now replicated in dozens of local authorities – see www.themertonrule.org). More recently, Uttlesford District Council in Essex has begun to use its powers as a planning authority to require consequential works on applications for major refurbishments and extensions by issuing a supplementary planning document (Uttlesford District Council). In the first year of operation (2006-2007) this policy had led to 485 planning approvals carrying the condition that consequential works be carried out. A survey of 72 sites, where work had started, showed that a total of 109 measures had been installed, and that only 20 out of the 72 required no consequential works: the other 52 all needed one or more of the measures to be installed (Energy Saving Trust, 2007). It is too early to tell how effective this policy might be, but it sets the very important precedent of making energy efficiency in the pre-existing dwelling a condition of planning approval. Devolved authorities in Wales and Scotland have also begun to diverge from Westminster in significant ways, while the Greater London Authority is active in this area with its Climate Change Agency, Green Concierge Service and energy-related policies. Such initiatives may lead to criticisms of patchy coverage and inconsistent standards, but it is nonetheless the case that a lead is being taken on the climate change agenda at the level where delivery has to take place – in local, regional and devolved government.

References

Allen, S. R., Hammond, G. and McManus, M.C. (2007) 'Prospects for and barriers to domestic micro-generation: a United Kingdom perspective', *Applied Energy*

Boardman, B. (2007) *Home Truths: A low-carbon strategy to reduce UK housing emissions by 80% by 2050*, Environmental Change Institute, University of Oxford

Boardman, B., Darby, S., Killip, G., Hinnells, M., Jardine, C.N., Palmer, J. and Sinden, G. (2005) *40% House*, 31, Oxford: Environmental Change Institute

Bordass, B. (2005) *Onto the Radar: How energy performance certification and benchmarking might work for non-domestic buildings in operation, using actual energy consumption, Discussion paper*

Chatterjee, E. (2007) *German Residential Housing Energy Reduction programmes*

Climate Outreach Information Network (2007) Last update, *ecovation*. Available from: http://ecovation.org.uk/ [November/28, 2007]

Darby, S. (2006) *The effectiveness of feedback on energy consumption: a review for DEFRA of the literature on metering, billing and direct displays*, Oxford: Environmental Change Institute

Department for Business Enterprise and Regulatory Reform (2006) *Our energy challenge: power from the people. Microgeneration strategy*, 06/993

Department for Environment, Food and Rural Affairs (2007) *UK Energy Efficiency Action Plan*

Department of Trade and Industry (2002) *Energy Consumption in the United Kingdom*

Department of Trade and Industry (2003) *Our energy future – creating a low-carbon economy*, London: The Stationery Office

Department of Trade and Industry (2006) *Construction statistics annual report 2006*

Energy Saving Trust (2007) *Energy in planning and building control – Uttlesford District Council's SPD on home extensions*, Energy Saving Trust

English Heritage (September 2007) *English Heritage interim guidance home information packs advice for domestic energy assessors*. Available from: http://www.english-heritage.org.uk/upload/doc/Advice-DEA2.doc [November/23, 2007]

Environmental Change Institute (2005) *40 percent house*. Available from: http://www.40percent.org.uk/

Environmental Change Institute (2007) *Reducing the environmental impact of housing – Appendix E Embodied energy*

Fawcett, T., Lane, K. and Boardman, B. (2000) *Lower carbon futures*, Oxford: Environmental Change Institute, University of Oxford

Hinnells, M. (2005) 'The cost of a 60% cut in CO_2 emissions from homes: what do experience curves tell us?' *British Institute of Energy Economics*, September 2005

Hinnells, M. (2006) 'Aiming at a 60% reduction in CO_2: implications for residential lights and appliances and micro-generation', *Energy Efficiency in Domestic Appliances and Lighting*, June 2006

House of Commons Communities and Local Government Committee (2008) *Existing housing and climate change*, HC 432-I. London: The Stationery Office

Janda, K. B. (2007) 'Turning solar consumers into solar citizens: strategies for wise energy use', *Proceedings of the ASES Annual Meeting*, 8-13 July 2007, American Solar Energy Society

Johnston, D. (2003) *A physically-based energy and carbon dioxide emission model of the UK housing stock*, Doctoral thesis, Leeds Metropolitan University

Johnston, D., Lowe, R. and Bell, M. (2005) 'An exploration of the technical feasibility of achieving CO_2 emission reductions in excess of 60% within the UK housing stock by the year 2050', *Energy Policy*, 33(13), pp. 1643-1659

Keesee, M. (2005) 'Setting A New Standard – The Zero Energy Home Experience In California', *ISES 2005 Proceedings*, International Solar Energy Society

Keirstead, J. (2007) 'Behavioural responses to photovoltaic systems in the UK domestic sector', *Energy Policy*, 35(8), pp. 4128-4141

Killip, G. (forthcoming) 'It's the size of the reduction target, stupid! The need for a wholesale re-think of energy efficiency policy in UK housing', *ACEEE Summer Study on Energy Efficiency in Buildings*, August 2008 forthcoming, American Council for an Energy Efficient Economy

Lowe, R. (2007) 'Technical options and strategies for decarbonizing UK housing', *Building Research & Information*, 35(4), pp. 412-425

Lowe, R. and Bell, M. (2000) 'Building Regulation and sustainable housing. Part 1: a critique of Part L of the Building Regulations 1995 for England and Wales', *Structural Survey*, 18(1), pp. 28-37

Olivier, D. (2001) *Building in ignorance. Demolishing complacency: improving the energy performance of 21st century homes*, Association for the Conservation of Energy/Energy Efficiency Advice Services for Oxfordshire

Poyzer, P. and Schalom, G. (2003) Last update, *Nottingham Ecohome*, Available from: http://www.msarch.co.uk/ecohome/ 23 October 2007

Roberts, D., Lowe, R. and Bell, M. (2005) 'Developing energy performance standards for UK housing: the Stamford Brook project', *The World Sustainable Building Conference*, 27-29 September 2005

Royal Borough of Kensington & Chelsea (2007) Last update, *flagship home*. Available from: http://www.rbkc.gov.uk/flagshiphome/general/default.asp [November/28, 2007]

Royal Commission on Environmental Pollution (2007) *The urban environment*, 26, London: The Stationery Office

Shorrock, L. and Utley, J. (2003) *Domestic energy fact file 2003*, Watford: BRE

Sustainable Energy Academy (2007) Last update, *old home super home alliance*. Available from: http://www.sustainable-energyacademy.org.uk/ [October 23, 2007]

Tirpak, D., Ashton, J., Dadi, Z., Gylvan Meira Filho, L., Metz, B., Parry, M., Schellnhuber, J., Seng Yap, K., Watson, R. and Wigley, T. (2005) *Report of the International Scientific Steering Committee International Symposium on the Stabilisation of greenhouse gas concentrations*, Exeter: Hadley Centre, Met Office

Uttlesford District Council (no date) *Energy efficiency condition*. Available from: http://www.uttlesford.gov.uk/climate+change/energy+efficiency+condition.htm [April 4, 2008]

CHAPTER 11:
Social housing –
'leading' the way?

Christoph Sinn, John Perry and Adrian Moran

Background

If we are to achieve 60 per cent or more in carbon reductions by 2050, it is absolutely vital that *all* sectors of the housing business rise to the challenge and do 'their bit'. As made clear in the previous chapter, most of these reductions will have to come from the existing stock, particularly owner-occupied and private rented properties, which make up over 80 per cent of the total UK stock. In fact the majority of energy-inefficient homes can be found in these sectors, illustrated by the relatively 'poor' average SAP ratings (Table 11.1).

Table 11.1: Average SAP rating by tenure 1996-2005

	Owner-occupied	Private rented	Local authority	RSL
1996	41.1	37.9	45.7	50.9
2001	44.4	41.9	49.6	56.4
2003	45.0	44.4	52.0	56.7
2004	45.6	45.7	53.9	57.3
2005	46.1	46.0	55.3	58.9

Source: CLG (2007).[1]

From this table it can also be seen that the social sector scores better. This is largely down to the decent homes programme but also due to the relatively young building stock. The requirement by the Housing Corporation on housing associations to achieve a certain EcoHomes rating as a pre-condition for funding will also have a bearing on the overall SAP score for the sector.

This brings us to an important point, namely that energy-efficiency improvements are to a great extent driven by regulatory requirements. Thus, the potential for considerable carbon savings in the social sector are huge, and it can be argued that social housing is in an ideal position to 'lead' the way in sustainable practices, as indeed it has been doing with the new-build agenda for some time (i.e. EcoHomes Excellent, now the Code for Sustainable Homes).

1   http://www.communities.gov.uk/documents/corporate/xls/statistics-2005

Whilst regulation is an important driver, another factor is that social landlords anticipate a long relationship with their housing stock. Unlike the owner-occupied sector with an average turnover of seven years, social landlords take a more long-term perspective, aligned to the typical 30-year business plan. Furthermore, there is the moral imperative to safeguard and improve the wellbeing of residents and communities, which also provides a rationale to engage with the climate change agenda. This is most evident in social landlords' efforts to tackle fuel poverty amongst residents.

This combination of working with often disadvantaged and vulnerable communities (thereby being in tune with those needs) and at the same time anticipating a long-term relationship with their housing stock, means that social landlords are naturally in a better position to anticipate future problems and take earlier action. As a result, much of the UK's research and development in the housing arena over the last 40 years has taken place in the social housing sector. Thus, it is not surprising that social housing, particularly that recently-built and managed by housing associations, is of a better quality than most of its private sector equivalent.

Given that climate change is now inevitable, future-proofing the stock becomes more and more important, as some of the extreme weather events seen in the last few years start to become a common feature. The floods in 2007, which had a devastating impact on a number of social landlords and their residents, particularly in the north, were a stark warning of things to come. Furthermore, with sea levels projected to rise by as much as half a metre or more by 2100 (IPCC, 2007), adaptation measures are urgently needed.

It has been suggested that some parts of the UK, such as the growth areas around Cambridge, could be under water by the end of this century (Wiles, 2007). It is thus vital that climate change considerations are a central part of any asset management strategy for existing stock, as well as being a key factor in developing new stock.

At the time of the first edition of this book (Bhatti et al., 1994), environmental issues had only just begun to appear on the sector's radar. This was largely due to the popularity of the concept of sustainable development, which gained widespread currency following the publication of the Brundlandt Report in 1987 (World Commission on Environment and Development, 1987). Environmentalism was still portrayed as some kind of 'sub culture', evoking images of 'tree-hugging' and 'hippiedom'. In housing terms it was entrepreneurial individuals and the self-build community who were paving the way (see Chapter 7). Apart from some trailblazing local authorities who signed up to the Local Agenda 21, not many housing organisations engaged with the agenda. Against this backdrop, the earlier book argued that there is a real need for a 'green housing policy'. Whilst we still fall short of a 'green housing policy', we have certainly succeeded in greening housing policy, as is evident in the raft of recent initiatives and targets referred to in this book.

Social housing, the Housing Corporation and the environmental agenda

Whilst climate change considerations have only started to enter the housing sector from the late 1990s onwards (see Chapter 7), other issues such as recycling and resource efficiency were starting to be addressed at earlier stages. A major drive to promote the environmental agenda around that time was provided by the Housing Corporation's Innovation and Good Practice programme. It helped to establish a number of projects and initiatives, whose impact can still be felt today.

The Housing Corporation's (HC) focus on the environment can be categorised into three main areas of activity – *projects*, *external*, and *internal*.

Projects

The very first Innovation and Good Practice (IGP) programme in 1996 had 'Housing Plus' as one of its key themes with the aim of promoting sustainable communities. The Corporation was particularly interested in projects that would develop resident and tenant involvement in local environmental projects. It was under this programme that the Corporation started to support the Sustainable Homes initiative by Hastoe Housing Association which has gone on to be a key source of information, promotion and help for associations and other organisations in implementing environmental activities (www.sustainablehomes.co.uk). One of the projects that the Corporation supported was the guide, *Developing an Environmental Policy and Action Plan* which supported housing associations new to this area in developing environmental strategies (Wayne, 2001).

Other projects included:

- the setting up of Sustainable Homes, a subsidiary of Hastoe Housing Association, which promotes awareness of environmental sustainability issues for housing associations;
- the Zero Waste and Zero Energy scheme by the BioRegional Development Group, which developed a concept design for a zero waste and zero fossil energy settlement in the Thames Gateway area;
- Sustainability Works, a web-based development tool for sustainable housing; and
- Routes to Sustainability, a comprehensive web resource to help housing organisations navigate through the sustainability maze.[2]

As well as supporting many individual projects on aspects of environmental sustainability over the years, such as energy and water, the Corporation chose

2   More projects overviews are available at http://www.housingcorp.gov.uk/server/show/nav.1418

Environmental Sustainability as one of the themes for the Gold Awards programme in 2007 to highlight and promote excellent examples of practice.[3]

External

Arguably, the biggest environmental impact that the Corporation has had to date has been through its investment policy. Working with the Building Research Establishment it was instrumental in developing the EcoHomes rating system which covers a number of environmental aspects, including energy, water and waste. Initially the Corporation expected schemes to reach EcoHomes *Good*. This was raised to *Very Good* for the 2006-08 National Affordable Homes Programme (NAHP). The 2008-11 NAHP round requires that homes funded should reach Code Level 3 (HC, 2007c). Requirements will increase over the coming years so that by 2016, all HC funded homes achieve Code Level 6, i.e. zero-carbon.

There are a number of other documents and publications that reflect the Corporation's approach to environmental issues, including the *Design and Quality Strategy* (HC, 2007a), *Design and Quality Standards* (HC, 2007b) and the Williams Report on the standards for homes in the Thames Gateway (Williams Commission, 2007).

It is now well recognised that a much bigger challenge is the carbon intensity of our existing stock. The Corporation anticipated this by commissioning the BRE to develop and pilot[4] EcoHomes XB (e**X**isting **B**uildings), an environmental assessment method for existing stock. Whilst this is not mandatory, it is promoted as good practice and supported by the Corporation through guidance and training (Sustainable Homes, 2007).

The Corporation's *Regulatory Code and Guidance* also makes reference to environmental sustainability. Under 3.4a it states that *'Permanent housing is sustainable, demonstrated by a commitment to effective protection of the environment and prudent use of natural resources'* (HC, 2005). The Corporation further expects environmental considerations to be part of associations' asset management strategies.

As Energy Performance Certificates (EPC) will be implemented for rented housing in October 2008, the Corporation commissioned Energy for Sustainable Development to work with housing associations to consider how EPCs will be implemented in the sector and to produce interim guidance (ESD, 2007).

---

3   Details of the three winners, Drum HA, Greenoak HA and Places for People Group are available at http://www.housingcorp.gov.uk/server/show/nav.3775
4   Two pilots were run to ensure the tool was appropriate for associations' operational requirements.

Internal

To demonstrate its commitment to the environment and sustainability, the Corporation published its *Sustainable Development Strategy* and action plan in June 2003 (HC, 2003). This was to feed into future corporate plans and reports on progress have been published (see for example HC, 2004). Part of the commitment was to have an internal 'Green' board which then steered the organisation to achieving the ISO 14001 Environmental Management Standard and accreditation, and revising the Corporation's procurement strategy.

The future

Practical guidance for social landlords on improving the environmental performance of their existing stock has only recently been published (HC, 2008). A key theme of the Corporation's 2008 *Innovation and Good Practice Programme* is 'Tackling Climate Change in Existing Stock' (HC, 2007d). The proposals received indicate that this may produce some ground-breaking research, which will inform future good practice in this area.

The key developments for the future are the establishment of the Homes and Communities Agency (HCA) and the Office for Tenants and Social Landlords (Oftenant). One of the HCA's objectives is *'…to contribute to the achievement of sustainable development'* (CLG, 2007), so it is likely that the environment and climate change will remain high on the agenda, particularly in the light of the zero-carbon targets and eco-town developments. Oftenant will have the power to set standards for registered providers within some constraints, including *'…the landlord's contribution to environmental, social and economic wellbeing of the areas in which their properties are situated'* (House of Commons, 2007).

Outlook

Despite the inroads made within the sector, there is no room for complacency, since the threat of irreversible climate change is here to stay for the foreseeable future. It is important to see environmental sustainability as a core business objective, rather than as an add-on. The following two case study chapters give a glimpse of the sectors' potential in taking on a clear leadership role to move us towards a low-carbon built environment.

References

Bhatti, M., Brooke, J. and Gibson, M. (1994) *Housing and the Environment: A new agenda*, Coventry: Chartered Institute of Housing

CLG (2007) *Delivering Housing and Regeneration: Communities England and the future of social housing regulation*, Communities and Local Government. Available from: http://www.communities.gov.uk/documents/housing/pdf/322429

ESD (2007) *Energy Performance Certificates: Interim Guidance for Housing Associations,* Housing Corporation. Available from: http://www.esd.co.uk/has/Good_Practice_Interim_Guidance_HAs.pdf

HC (2003) *Sustainable Development Strategy,* Housing Corporation. Available from: http://www.housingcorp.gov.uk/upload/pdf/susdevstrategy.pdf

HC (2004) *Sustainable Development Report 2004,* Housing Corporation. Available from: http://www.housingcorp.gov.uk/upload/pdf/Sustain_Develop04.pdf

HC (2005) *The Regulatory Code and Guidance,* Housing Corporation. Available from: http://www.housingcorp.gov.uk/upload/pdf/RegulatoryCode.pdf

HC (2007a) *Design and Quality Strategy,* Housing Corporation. Available from: http://www.housingcorp.gov.uk/server/show/ConWebDoc.10782

HC (2007b) *Design and Quality Standards,* Housing Corporation. Available from: http://www.housingcorp.gov.uk/server/show/ConWebDoc.10783

HC (2007c) *National Affordable Housing Programme 2008-11 Prospectus,* Housing Corporation. Available from: http://www.housingcorp.gov.uk/server/show/ConWebDoc.11921

HC (2007d) *Innovation and Good Practice Programme Prospectus 2008-10,* Housing Corporation. Available from: http://www.housingcorp.gov.uk/upload/pdf/IGP_08-10_prospectus.pdf

HC (2008) *Fit for the Future: The green homes retrofit manual,* Housing Corporation.

House of Commons (2007) *Housing and Regeneration Bill.* Available from: http://services.parliament.uk/bills/2007-08/housingandregeneration.html

IPCC (2007) *Climate Change 2007: Summary for Policymakers*, Intergovernmental Panel on Climate Change. Available from: http://www.ipcc.ch/pdf/assessment-report/ar4/syr/ar4_syr_spm.pdf

Sustainable Homes (2007) *EcoHomes XB: A guide to the EcoHomes methodology for existing buildings*, Housing Corporation. Available from: http://www.housingcorp.gov.uk/upload/pdf/EcoHomes-XB.pdf

Wayne, J. (2001) *Developing an Environmental Policy and Action Plan: A Guide for Housing Associations*, Sustainable Homes. Available from: http://www.sustainablehomes.co.uk/pdf/ENV%20policy.pdf

Wiles, C. (2007) 'Are these the homes of the future?', *Inside Housing*, 9 February

Williams Commission (2007) *The Williams Report – Quality first: The Commission on the Design of Affordable Housing in the Thames Gateway*, Housing Corporation. Available from: http://www.housingcorp.gov.uk/upload/pdf/Thames_Gateway_final.pdf

World Commission on Environment and Development (1987) *Our Common Future*, Oxford: Oxford University Press

CHAPTER 12:
Case study:
A changing environment – creating sustainable housing for the future: Places for People

Nicholas Doyle

Energy will become one of the defining issues of the 21st Century. Its relationship to climate change and the impact of peak oil will raise issues about its sources, uses, impacts, costs and ownership. Housing – the homes we build and the communities we create – will feel the full impact of this but also has the potential to be central to a new paradigm that moves us beyond carbon. Already, energy is rising up the social, political and economic agenda, and transforming the way we see the world (Bulkeley and Betsill, 2003). If we take an optimistic view, action on both climate change and peak oil are mutually compatible and have led to a range of policy and strategy initiatives, including the Climate Change Bill (House of Lords, 2007) and the Energy White Paper (DTI, 2007).

Our increasing demand for energy and the finite supply of oil are changing the traditional way we view our lifestyles. Rises in the global cost of energy have seen UK households experience a 50 per cent increase in domestic energy costs in the last two years.

This has seen 1.7 million more people plunged into fuel poverty over that period, meaning there are now over three million people spending over ten per cent of their annual income on energy bills – a figure which will rise if the effects of climate change are realised as oil runs out and prices rise accordingly (EEPH, 2007). As a consequence, some commentators predict an impending global energy supply crisis, while others such as the US Department for Energy have called the peaking of world oil production, *'...an unprecedented risk management problem, where fuel prices and price volatility will increase dramatically'* (Hirsch, 2005). Deal III (2005) states that with other nations increasing their use of oil as an energy resource, energy prices will continue to increase and everyday supplies become more and more limited. With 95 per cent of the goods we consume relying directly or indirectly on oil, the knock-on effect will be felt by millions of people with higher food, transport, and energy costs, as well as higher inflation.

A changing landscape

The UK's 22 million households will be among those most affected by the threat of climate change and the finite supply of oil. Over a quarter of the UK's carbon emissions are produced by the energy we use to heat, light and power our homes, over twice the amount produced by the nation's fleet of 32 million cars (Liberal Democrats, 2007). Combating climate change starts at home, literally, and is now at the forefront of political, social and economic policy. In 2006 the government issued three key documents that outlined its approach to securing sustainable development and a carbon neutral landscape. With the publication of *Building a Greener Future: Towards Zero Carbon Development*, the new Code for Sustainable Homes, and the Planning Policy Statement, *Planning for Climate Change*, it is clear that the housing landscape will be changed forever, with new ways of planning, designing, building and powering our homes needed in a carbon neutral environment.

The impact of climate change on towns and cities will also be significant. Temperatures will rise by up to five degrees; there will be hotter and drier summers, and wetter, windier winters. There will be greater incidences of flooding – sea levels will continue to rise for several centuries. By 2090 it is estimated that sea levels could rise by up to 60cm, severely impacting on existing housing (Shaw *et al.*, 2007). Around two million UK properties could be at risk from flooding and, according to the Environment Agency around 500,000 are at significant risk, a figure that has doubled in the last seven years (Environment Agency, 2005).

Increasingly, government and the housing industry are realising that a twin track approach of adapting to and mitigating against the threats posed by climate change is vital if we are to continue offering sustainable and vibrant communities in places where people want to live, work and play. Providing decent homes is intimately bound up with the question of climate change and recognising this link '...is essential for social cohesion, personal wellbeing and self-dependence' (Edwards and Turrent, 2000, p.12).

Against this backdrop is the pressure of increasing housing supply. In the last 30 years there has been a 30 per cent increase in the number of households, yet a 50 per cent drop in the level of housebuilding. Government now wants the housebuilding programme to deliver three million new homes by 2020, providing the housing and construction industry with an important opportunity to build high standards of sustainability into the UK's future housing stock.

While the business sector adapts its practices and holds itself accountable for the environmental impact of its activities, Smales (2007) argues that the housing industry fails to take a long-term view on sustainability, stating the '...*housing industry is prone to take the shortest and easiest route to the maximum profit'*.

Yet, in contrast to Smales' view, there is now a growing groundswell of action and opinion that the housing and construction sector should tackle the threats posed by climate change. Climate change is here and now, and the industry is already beginning to adapt to and mitigate against the effects, pushed by government regulation and policy. There are already some very good examples of building sustainably, but if we are going to take these from the margins to the mainstream, we need to become much better at sharing best practice and information within the sector. We must also cut through the environmental 'white noise' – the never-ending debate about details and options – if we are to stop the paralysis of indecision that is preventing the sector taking action.

As a first step, the energy hierarchy – saving energy before changing its source – must now become part of everyday housing vocabulary. Our aim should be to drive down our need for energy. This will reduce costs and emissions and then allow us to look at issues around where we get energy from and what kind it is. This will include implementing intelligent sustainable design to improve the fabric of new and existing homes and the use of renewable and community energy. A useful way to conceive of these measures is to categorise them in terms of place, power and people which enables industry to implement a holistic model in combating the effects of climate change.

At Places for People – one of the largest property management and development organisations – we have begun to 'mainstream' sustainability through all areas of the group's activities, from delivering efficient new-build housing; improving the performance of existing homes; delivering new ways of generating power and energy and challenging people's perceptions and attitudes as to how they consume electricity and gas. It was an approach recognised by the housing sector, when the group recently picked up the Housing Corporation's Gold Award for Environmental Sustainability.

Place – creating sustainable communities

The housing industry has a golden opportunity not only to meet the environmental challenges laid down by government and meet housing targets, but also to create places where people are proud to live and work and are truly sustainable.

One example of our work is Broughton Square – a £27 million development of 229 homes set on the eastern edge of Milton Keynes. The development comprises properties for private sale, shared ownership, affordable rent and intermediate rent for key workers. With Broughton Square, Places for People has demonstrated to the industry that sustainable housing can be delivered cost-effectively on a large scale. All the homes have met the former EcoHomes 'excellent' standard, which came at an additional cost of £1,880 per property – significantly lower than was previously possible at the time.

The task was to build an attractive, large-scale, modern development, with excellent environmental credentials, on a reasonable cost-base. The finished product accomplished this, setting a benchmark standard for affordable, desirable and sustainable homes. Providing high standards of environmental performance on a large scale, without running up hazardous costs, was always going to be difficult. To do this, the group involved an EcoHomes assessor at the start of the design and build process. The result of this was a clear focus on sustainability throughout the design and planning of the properties. It was more effective to fit properties with those features which would improve their EcoHomes ratings early on, rather than design the properties independently, and add-in these features later. In addition, the homes have boilers with the lowest possible levels of nitrogen oxide emissions. A Sustainable Urban Drainage System has also been incorporated into the design to handle the discharge of surface water and encourage water attenuation on site, whilst helping prevent pollutants being washed into rivers and groundwater supplies.

To have this emphasis on sustainability in the design of new homes brings with it the danger that the homes may look too futuristic and be unattractive to prospective residents. The design at Broughton Square, however, exemplifies how newly built houses can be designed for environmental performance and yet retain traditional appearances. So in addition to being environmentally friendly, the homes are genuinely desirable dwellings.

For all of Broughton's environmental credentials, there are some issues that the development does not address. To start with, if all new homes are to be zero-carbon by 2016, and if existing homes are to have their carbon emissions reduced in line with government targets for 2050, the housebuilding sector, the government and energy companies need to work to provide cost-effective, renewable sources of power. Places for People is currently working towards this in many of its other developments, with the inclusion of wind turbines, solar panels, and ground-source heat pumps.

Yet environmental sustainability in the housing sector is not just about new-build housing and the race to develop zero-carbon homes. If we rely on delivering these homes we will only save five per cent of our carbon emissions by 2050 when the real target is at least 60 per cent. Improving the energy efficiency of the UK's existing housing stock will have a far greater impact on reductions in carbon dioxide emissions. Over 500,000 existing homes would have to be refurbished every year from now until 2050 if we are to meet the government's long-term target of cutting carbon emissions by 60 per cent. This is a monumental task and already the housing industry is failing to keep pace.

The housing sector urgently needs to develop an overarching environmental strategy that focuses on existing stock as well as new homes. Once existing homes have the basic energy-efficiency measures installed, such as loft and cavity wall insulation, the limitations – because of the existing fabric, impact on the community and neighbours, planning and the cost of refurbishment – shift the emphasis onto energy use and to a lesser extent energy production. Put in simple terms the options are far more limited and the costs much higher. It is no surprise therefore that much of the emphasis on improving energy efficiency has been on new build.

With a managed stock of around 60,000 homes, Places for People is beginning to take the necessary steps and improve the thermal efficiency of our properties and has already demonstrated how significant steps can be made to raise the standard of existing homes far above the national average. Through our 'Affordable Warmth' strategy we have invested millions of pounds in our homes, and have delivered energy bill savings for our customers of up to £200 year. This has seen our thousands of homes receive an average 'SAP' rating of 70.

Working in partnership with utilities we have also insulated 6,500 homes at a cost of over £3 million; carried out cavity-wall and loft insulation work; and have provided over 300,000 free low-energy light bulbs to residents. This inititiative has helped over 100,000 customers save £13.4 million pounds in energy costs, whilst reducing carbon dioxide emissions by over 46,000 tonnes. Furthermore, the group secured a £15.6 million grant from the Big Lottery Fund to regenerate over 75 local communities across the UK to create green community spaces, and recreational activities for thousands of their customers. It is one of the biggest grants given out by the Big Lottery Fund and will make a significant contribution to 300,000 people's lives, helping to provide access to high-quality green space, strengthening community spirit and reducing the impact of climate change.

The project recognizes that good quality green spaces are essential to the wellbeing and future of our towns and cities. A well-designed environment can transform local communities, providing places for people to play and relax. They can also have a positive impact on health, can help tackle antisocial behaviour, increase biodiversity, and help reduce the impact of climate change.

Powering the homes of the future

The UK renewables industry has had an enviable reputation at various stages in the past and has perhaps the widest choice of renewable energy sources of any country in the world, having wind, water, solar, biofuels, geothermal energy, tidal and wave energy resources to hand. These technologies have led the government to set a target of having 20 per cent of the UK's energy coming from renewable sources by 2020. Currently, less than five per cent of our energy comes form renewable sources with only four in 1,000 homes having any kind of low- or zero-carbon energy source (Boardman, 2007).

This is the time for business and the housing industry to show leadership, and help meet the target by investing in the renewables industry. If we do not, the impact will be poor quality renewables which are low in availability, and fall drastically short of the government's targets. Places for People has recently begun working in partnership with Viridian Solar to install new, low-cost solar panels in 250 of its homes across the country, the first of which have been installed in Walker Riverside – a market renewal project in Newcastle. By mainstreaming the technology into its building programme, the solar panel units cost below £1,000 per household – around a quarter the cost of traditional solar panel systems – making them far more affordable for new housing developments and existing homes. The solar panels provide up to 60 per cent of the annual hot water demand for the homes to which they are fitted.

Places for People is also trialling wind turbines at its development in Wolverton, Milton Keynes and with existing housing in Nottingham. The £50 million Wolverton development will see 300 new homes built on a historic site, once home to the Royal Train Shed. If the wind turbine trials are successful, Places for People will install a further six. On top of this the company has installed solar heating panels to provide domestic hot water. In combination, these features should provide ten per cent of the energy demand of the development.

Places for People has also been trialling the use of combined heat and power units which produce both heat and electricity at the same time, at two schemes in Newcastle – Stanhope Street and Clayton Street. In its simplest form, these units employ a gas turbine, an engine or a steam turbine to drive an alternator and the resulting electricity can be used either wholly or partially on-site. When fuel is burnt to produce electricity, heat is generated. This heat is then captured and utilised for some useful purpose such as space heating, water heating or refrigeration. Because CHP systems make extensive use of the heat produced during the electricity generation process, they can achieve overall efficiencies in excess of 70 per cent at the point of use, and help reduce carbon dioxide emissions by up to 15 per cent. The two schemes which have nearly 600 properties between them, have helped over 1,200 residents save hundreds of pounds on their energy bills. The new system at Stanhope Street has cut energy use on the estate by an estimated 64 per cent. Weekly heating bills now average just £3.35 per dwelling and the area has seen a 4,200 tonne reduction in carbon emissions.

The Group is also exploring the use of and installing ground-source heat pumps – a heating system that uses the earth's core to heat warm water, and can help reduce energy bills by up to 60 per cent. It has installed ground-source heat pumps at Arbury Park, Cambridge in homes for affordable rent. Ground-source heat pumps provide a new and clean way of heating houses. The system is a simple one that transfers the heat in the ground into the home's heating system. As it is not burning fuel, it reduces the impact on the environment by producing fewer CO_2 emissions and is three to four times more energy efficient than a gas boiler.

Finally, Places for People also looks to provide an element of investment in tackling climate change in developing countries. It has provided funding to the 'Proyecto Sol' (Project Sun) scheme in Nicaragua, described in Chapter 17, which installs solar panels for electricity production in the homes of low-income farming families. Its latest tranche of funding, in 2008, will see a total of well over 100 families included in Proyecto Sol, and a sustainable revolving fund established which means that up to 20 further households can be brought into the scheme each year in future.

People

However, if we are to mainstream environmental sustainability then we need to realise that it is people who will drive forward, and deliver, long-term environmental change. Encouraging customers to think about the energy they use, its cost and impact on the environment, can have a dramatic and positive effect on people's behaviour, while at the same time helping to reduce the effects of climate change.

Places for People has begun installing display meters at a number of its developments. These simple devices show residents how much electricity they are using, how much it is costing, and how much harmful CO_2 they are emitting. This has encouraged residents to think about how much energy they are wasting, with residents recording an average saving of 13 per cent.

The introduction of Energy Performance Certificates for the sale or rent of homes, despite their difficult birth and high cost in the UK, will begin to empower householders to make informed decisions about the homes they live in. The energy performance of homes will become as common a currency as miles per gallon are for cars. Furthermore, the housing and construction industry must urgently begin to invest in people delivering the new green environmental landscape. It must invest in the skills and knowledge of the people that will create the communities of tomorrow and not shy away. Our industry can make a major and positive contribution by promoting policies which increase energy efficiency in new build and existing developments and stimulate planning and development programmes to meet housing targets in a green and efficient manner.

The housing industry's approach to environmental sustainability must be long-term and focused on reduced costs for customers and the use of technologies or practices which minimise the effects of climate change. This will inevitably mean finding new ways to reduce energy consumption and our production of waste, it will mean researching and using new techniques and technologies, and it will mean finding new ways to use existing funding to do much more.

After years of debate and discussion, the future is now clear – environmental action can no longer be the preserve of the green elite or confined to one-off projects. The challenge for all of us is how we get the climate change agenda, once and for all, to move from the margins to the mainstream.

References

Barker, K. (2004) *Review of Housing Supply – Delivering Stability: Securing our Future Housing Needs*, HM Treasury. Available from: http://www.hm-treasury.gov.uk/consultations_and_legislation/barker/consult_barker_index.cfm

Bulkeley, H. and Betsill, M. (2003) *Cities and Climate Change: Urban sustainability and global environmental governance*, London: Routledge

Boardman, B. (2007) *Home Truths: A low-carbon strategy to reduce UK housing emissions by 80% in 2050*, Environmental Change Institute, University of Oxford. Available from: http://www.eci.ox.ac.uk/research/energy/hometruths.php

Deal III, W. F. (2005) 'The Magic of Energy: Oil Surely Changed the Way That Humans Value Energy', *The Technology Teacher*, Vol. 65, p10

DTI (2007) *Meeting the Energy Challenge: A White Paper on Energy*, Department of Trade and Industry

Edwards, B. and Turrent, E. (2000) *Sustainable housing principles and practice*, London: Routledge.

EEPH (2007) *The impact of rising fuel prices in managed housing*, Energy Efficiency Partnership for Homes. Available from: http://www.eeph.org.uk/uploads/documents/partnership/FinalSummaryReportD3Final%20_3_.pdf

Environment Agency (2005) *The Foresight Future Flooding Report*. Available from: http://www.environment-agency.gov.uk/subjects/flood/763964/?version=1&lang=_e

Hirsch, R. (2005) *Peaking of World Oil Production: Impacts, Mitigation, and Risk Management*, US Department of Energy. Available from: http://www.netl.doe.gov/publications/others/pdf/Oil_Peaking_NETL.pdf

House of Lords (2007) *Climate Change Bill*, London: TSO

IPCC (2007) *Climate Change 2007: Summary for Policymakers, Intergovernmental Panel on Climate Change*. Available from: http://www.ipcc.ch/pdf/assessment-report/ar4/syr/ar4_syr_spm.pdf

Liberal Democrats (2007) *Zero carbon Britain: Taking a global lead*. Available from: http://www.libdems.org.uk/media/documents/policies/zero%20carbon.pdf

Shaw, R., Colley, M. and Connell, R. (2007) *Climate Change Adaptation by Design: A guide for Sustainable Communities*, Town and Country Planning Association. Available from: http://www.tcpa.org.uk/downloads/20070523_CCA_lowres.pdf

Smales, J. (2007) 'Why developers just don't get it', *Regenerate Live*, October. Available from: http://www.regeneratelive.co.uk/story.asp?sectioncode=697&storycode=3098042

CHAPTER 13:
Case study:
Black Country Housing Group

Richard Baines

This short chapter gives an overview of the various environmental projects and initiatives Black Country Housing Group (BCHG) have been developing over recent years. Early on the group realised the importance of incorporating environmental principles within their operations. BCHG first introduced their comprehensive environmental policy in 1999, covering all areas of their business. They can be fairly described as one of the trailblazers within the social housing sector.

Whilst responsibility for actively implementing the policy lies with each departmental director, the overall lead on BCHG's sustainability work is taken by 'e²S', the group's in-house environmental consultancy. e²S provides environmental consultancy services to BCHG, other social landlords, local authorities, designers, engineers, cost consultants and manufacturers. It also undertakes surveys, develops general and detailed designs, presents physical and virtual models and monitors actual performance. e²S has pioneered a 'green' design process which can be applied to any project on any site (Box 13.1).

Box 13.1: The *green* design process

- Consultation with a wide range of stakeholders, including tenants' representatives, local residents, specialist suppliers and sub-contractors, local authorities, government departments, research and housing staff.
- Advanced computer simulations that predict future performance of buildings.
- 'Green' building specifications.
- Environmental impact assessment, from 'cradle to grave'.

In 2005, BCHG produced a revised environmental policy, setting out how it intended to minimise the environmental impact of its future operations. The policy looks at all business areas, from office services through to housing/assets management and development. The well-known sustainable development paradigm of promoting *'...development that meets the need of the present without compromising the ability of the future generations to meet their own needs'* (WCED, 1987) was the guiding principle of the policy.

An annual review is carried out in order to incorporate any new environmental developments and innovations made by the group, which feeds into one overarching sustainability strategy to include economic, social and environmental policies. This enables BCHG to consider actions as a whole. The effectiveness of the sustainability strategy is then put through a vigorous audit process that assesses overall progress toward set targets (Box 13.2).

Box 13.2: BCHG Environmental Audit (summary)

Commitment:
- full support obtained from departmental directors;
- communicating commitment to all staff.

Objectives:
- verification of legislative and regulatory compliance;
- assessment of internal policy;
- establishing a 'baseline' of current practise status;
- identification of improvement opportunities.

Targets:
- material management;
- energy management and savings;
- water management and economy of use;
- waste generation, management and disposal;
- transportation and travelling practices;
- staff awareness, participation and training;
- preservation of historic countryside and townscape.

Process:
- on- and off-site audit;
- collating of data and analysis;
- review of audit report;
- reporting to directorate.

Follow up:
- action plan;
- arranging next audit scope and schedule.

Environmental activities

The following pages provide an overview to the group's environmental activities, since the mid-1990s. It is not meant to give detailed technical information about each initiative. For this, interested readers are encouraged to contact e²S directly for more information and advice.[1]

---

1   www.bcha.co.uk, Tel. 0121 561 1969.

Green Futures

Green Futures was initiated 1994, just two years after the Rio Conference on sustainable development. It was originally supported by the European Regional Development Fund (ERDF) and went through various phases (Box 13.3).

Box 13.3: Green Futures

1995 – 1997	Initial research
1997	15 'control' houses constructed at Bryce Road
1998 – 2000	Further research
1998 – 2002	Monitoring all properties' performance
1999 – 2004	12 'experimental' houses and the subterranean Eco-Pod constructed at Bryce Road

---

Green Futures incorporates the principles of sustainable development, striking a balance between economic, environmental and social benefits, such as:

- greatly reduced space heating bills (from as little as £35 per year);
- 50 per cent saving on hot water heating;
- the potential for carbon-neutral or carbon-zero power supply;
- near zero-carbon emission heating;
- year-round thermal comfort;
- 30-50 per cent savings on drinking water consumption;
- elimination of flood risk due to storm-water run-off;
- over 90 per cent of materials (by weight) sourced within the West Midlands; and
- development and demonstration of five new products with key environmental benefits, all locally manufactured within a 20 miles radius (Box 13.4).

Box 13.4: Green Futures product innovations

SureStop.	remote water stop tap
SkyTube	brings light into the house
Corium	thin brick cladding system
Chamois Eco Kitchens	made of recycled wood
Permadoor.	doors using 98 per cent recycled plastic

---

Funding: Advantage West Midlands, Black Country Housing Group, ERDF, Housing Corporation, private investment.

Bryce Road and the Eco-Pod (1997-2004)

Part of Green Futures, the Bryce Road (brownfield) development, comprising 27 homes, is one of BCHG's flag-ship scheme. The scheme incorporates systems that conserve non-renewable and dwindling resources and greatly reduce waste and pollution, both in building construction and in use. Winning the Homes Award for Innovation in 2000, it is now a national demonstration project. The £1.5 million development also features the UK's first Eco-Pod, a 230 square metre basement where visitors can study the design, fabrication and construction of the houses. Computer graphics, video records, products and displays support training and provide a unique insight into the reality of sustainable development.

Energy consumption is minimised by 'super insulation' – very high standards of thermal insulation and airtight construction with managed ventilation. The scheme makes maximum use of pre-assembled building components, including pre-fabricated walls, floors and roofs and factory-produced kitchen and bathroom pods.

The main objectives of the project are to:

- demonstrate a range of green technologies to assess and compare their performance, cost, energy consumption and environmental impact;
- monitor various green technologies to assess and compare their performance, cost, energy consumption, and environmental impact;
- consult with residents on their experiences of living with green technologies;
- address the apparent lack of 'eco-technological' knowledge in the construction industry at the time;
- use local products and assist local manufacturers to develop and trial new products; and
- maximise 'off-site' manufacturing and pre-assembly of building components to improve quality control, cut waste and reduce construction time on site.

Various green technologies are employed to minimise the environmental impact of the development, such as:

- alternative heating and power systems, including solar (photovoltaic) panels, solar water heaters, wind power, sun tubes to maximise 'natural' lighting, ground source heat pump and a gas-powered combined heat and power (CHP) unit to generate electricity and provide hot water;
- advanced multi-layer insulation, including special glazing systems;
- water conservation devices ('grey' water from baths and washbasins and harvested rain water, cleaned and stored for use in toilets and washing machines);
- waste recycling, such as an odourless composting toilet and units to segregate rubbish at source; and
- storm water remediation techniques, such as porous pavement and rain water harvesting systems.

The scheme was developed in partnership with a number of agencies, including Dudley Metropolitan Council, Advantage West Midlands and Birmingham City University (formerly University of Central England).

Funding: Dudley Metropolitan Council, ERDF and Single Regeneration Budget.

Vantage Point

Vantage Point on the Lyng Estate in West Bromwich is a four storey block of apartments designed to provide accommodation for the elderly, comprising 21 single- and 22 two-bedroom apartments as well as apartments for visitors and staff.

The brief was to reduce the amount of fuel used and to provide a high standard of comfort to the occupants. The scheme incorporated many of the same technologies as the Bryce Road development to achieve high environmental standards, such as:

- super insulation techniques, which cut fuel use by 50 per cent;
- under-floor heating, which removed the need for radiators and resulted in substantially reduced running costs;
- thermal mass, combined with zone control of the heating system to prevent overheating; and
- a small-scale CHP unit, which proves to be 200 per cent more efficient at turning natural gas into electricity compared to a traditional power station.

Funding: Capital Challenge, Housing Forum, Sandwell Metropolitan Council and BCHG.

Norwood Road

The Norwood Road scheme was designed to tackle fuel poverty in 22 council-owned houses in Dudley. The properties were traditional, single-skin brick walled dwellings, built between1928-43; with a SAP rating of zero. Most were heated by gas fires and supplementary electric heaters, with some only using direct-acting electric heaters. A year-long post-completion study was carried out in the homes by e^2S, on behalf of Dudley Council, with the aim of monitoring the effect of several new measures to increase energy efficiency of gas and electricity consumption.

The total predicted fuel bills fell from an average of £1,021 per annum to just £135 for heating and hot water. It could be shown that just £350 worth of insulation reduced the heat lost through the walls by 90 per cent, more than enough to achieve savings of over 60 per cent.

To achieve this, the houses were insulated externally (at a cost of £6,000 each), fitted with micro-boilers and a small number of radiators making the homes comfortable without the need for central heating. Passive stack ventilation was retro-fitted and those homes with south-facing roofs received solar water heaters. On other properties, either grey water recycling or rain water harvesting systems were installed. One wind-powered street light was set up to raise general environmental awareness. Window frames, kitchen units and internal joinery were reused to further reduce the environmental impact. Wiring and insulation aimed to take into account future needs.

Funding: Housing Corporation, Dudley Metropolitan Council, BCHG.

The hydrogen fuel cell projects

The hydrogen fuel cell projects are an experiment with the technology in pilot properties run by BCHG; now in conjunction with the University of Birmingham and BAXI Heating Group and funded by Advantage West Midlands.

Hydrogen is *the* fundamental chemical element that caries energy in fuels. For example, natural gas comprises four hydrogen atoms attached to one carbon atom. It is easy to break the bonds between the hydrogen and carbon by applying a little heat. The bond-breaking process gives off energy and the process is self-sustaining. Unfortunately the carbon released combines readily with oxygen to form carbon dioxide, the gas responsible for global climate change.

Fuel cells convert hydrogen and oxygen into water. In the process a spare electron is released and this can form an electrical current. The process also gives off heat. Fuel cells can be carbon-free in operation. If future fuel cells are manufactured by processes that are themselves powered by carbon-free energy (e.g. from other fuel cells), we could single-handedly resolve the issues of energy-induced climate change and energy/fuel security.

The projects consist of two phases. Phase one saw the establishment of the UK's first fuel cell trial system in the domestic sector. The first of the pilot properties was to provide baseline information leading to performance-enhancing modifications to a second property in the trials. In phase two, BCHG signed up to a partnership which is exploring alternatives to traditional methods of generating domestic heat and power, including the new prototype BETA fuel cell system (Plate 13.1).

The successful utilisation of fuel cell technology in the domestic sector would be a huge step forward in creating a low-carbon society. The benefits include:

- carbon-free energy systems;
- resolving the sustainability of energy supply in the UK's housing stock and beyond;
- addressing the issue of energy (in)security; and
- eliminating fuel poverty.

Funding: Department of Trade and Industry (phase one), Advantage West Midlands (phase two), BCHG, Venture Capital, University of Birmingham.

Plate 13.1: Latest installed fuel cell system

Source: BCHG.

Sandwell Solar City

This project aims to overcome the high cost and slow payback associated with solar photovoltaic panels that convert light into electricity.

The conventional methods of generating electricity in the UK are unacceptably inefficient; at best wasting 65 per cent of the fuel used at the power station.

The Sandwell Solar City scheme was conceived to address this and associated issues, namely:

- security of energy supply;
- creating market-demand for solar technology;
- establishing the West Midlands as a manufacturing centre for solar technology;
- encouraging private sector investment; and
- creating new job opportunities for local people.

The idea is to 'rent' roof space and install photovoltaic cells (over 20,000 m² of PV panels) to generate electricity on a city-wide scale and at power-station intensities. The district network will be reconfigured to accommodate this embedded generation and the household or business underneath just buys and uses electricity, oblivious to how it is generated. Economies of scale, coupled with the availability of new, thin-film PV cells, which are half the cost of conventional cells, would make this scheme feasible. Still at the conceptual stage, the project would see the coming together of BCHG, Sandwell Metropolitan Borough Council, a number of housing associations, private developers, as well as academic and research institutions.

Funding: Communities and Local Government through the Housing Market Renewal Pathfinder programme (Urban Living), various public and private investments.

Changing Rooms That Don't Cost the Earth

This project was implemented in partnership with Urban Living and Jephson Housing Association, and involved building a showcase house to demonstrate how homes could be both stylishly and sustainably renovated. Several houses opened to the public from March 2004. Subsequently, e²S offered a consultation and advice service for those home owners who potentially, through viewing the showcase, were interested in finding out more about renovating their own homes without damaging the environment.

The overall aim of the project was to look at:

- the application of low environmental impact technology (Box 13.5);
- showcasing high-level design and material quality standards to generate interest in sustainable refurbishment;
- encouraging home owners with the available resources to think about the environment when refurbishing their homes; and
- thereby stemming the abandonment of older towns.

Funding: Urban Living.

Box 13.5: Changing Rooms features

Warm roof:
- high insulation levels;
- loft converted into bedrooms;
- roof windows;
- solar panels.

Draught proofing and ventilation:
- retained window frames;
- upgraded double glazing;
- draughtproofed lobbies;
- thermal insulation;
- airtight construction.

→

Kitchen:
- fuel-efficient boiler and controls;
- hot water cylinder and pipe insulation;
- energy-efficient domestic appliances;
- recycling facilities.

Bathroom:
- dual flush toilets;
- spray taps;
- low-flow shower.

Other features:
- recycled building materials;
- super insulation;
- low-energy lighting in all rooms;
- solar water heating;
- heating bills reduced by 75 per cent;
- CO_2 emissions cut by over 80 per cent.

Energyextra

Energyextra is an Energy Services Company (ESCo), set up in partnership with a number of other housing associations, namely BCHG, Accord HA, Jephson HA, South Staffordshire HA, Gloucestershire HA, Home Zone and Dudley Metropolitan Borough Council. Membership is available to tenants and service users of each partner. Participation means easy access to cheaper gas and electricity through a preferred supplier partnership as well as help and advice for improving energy efficiency in the home.

The principal aim of the project is to alleviate fuel poverty in domestic properties, by using a combination of services including cheaper energy supplies, targeted face-to-face energy efficiency and welfare benefit eligibility advice to maximise household income.

The initiative delivered to residents:

- savings of £58-£158 per year through energy-efficiency improvements;
- access to cheaper gas and electricity through fuel supply partner;
- discounts on energy-efficient white goods;
- free low-energy light bulbs; and
- insulation/heating improvements utilising the funding available through the Energy Efficiency Commitment (EEC).

Funding: After initial funding from the Energy Saving Trust for a pilot feasibility study, funding for the day-to-day operation of Energyextra comes from the void referral fees from the preferred fuel supply partner.

References

World Commission on Environment and Development (1987) *Our Common Future*, Oxford: Oxford University Press

CHAPTER 14:
'Greening' the housing market – putting the green housing market in context

Adrian Wyatt

The UK housing market has been characterised for several decades by rising demand, generally rising prices and a shortfall in supply. It is therefore arguable that the recent government drive to 'green' the housing market could not have come at a worse time – adding both cost and complexity in a situation where the last thing the government wants to do is to provoke slower delivery and further price increases. This chapter sets out to review the recent trends and developments in the housing market from a 'green' perspective, and assess the prospects for a significant shift towards a 'green' market, acknowledging the difficult conditions for radical change.

Basic housing demand

Without doubt, the UK's housing market is in the midst of rapid and fundamental change. As highlighted in the Barker Report (2004), demand for housing had been increasing, driven primarily by demographic trends and rising incomes. However, by 2001, construction of new houses had fallen to its lowest level since the Second World War, with these factors combining to produce an estimated trend rate for UK real house price growth over the last 30 years of 2.4 per cent, considerably higher than the European average of 1.1 per cent for the same period.

Taking the level of private sector build into account (starts and completions) for 2002-03, it was estimated that a reduction in the trend in real house prices to 1.8 per cent would require the construction of an additional 70,000 private sector homes each year, raising private sector construction output to around 210,000 homes per annum. In addition, the Barker Report estimated that an increase of about 26,000 new social housing starts would be needed to take account of demographic projections and to make inroads into the backlog of unmet need. Yet in 2002-03, the number of social houses built in the UK was only 21,000.

By 2007, the Housing Green Paper *Homes for the Future: More affordable, more sustainable* estimated that 223,000 more households are being formed each year and that an output of 240,000 new homes would be required each year by 2016.

In short, the UK market has been characterised by rising demand and rising prices, driven by a shortfall in supply, and conditions (despite the dip in the market in 2008) appear to be worsening, not getting better.

Against this background, the government has declared its goal of mitigating the negative impact of the built environment on climate change, focusing its attention particularly on newly built housing.

The built environment, environmental change and key environmental legislation

Before looking at what the drive towards more environmentally friendly building might actually require, it is worth reminding ourselves of a few key facts which are driving this wave of well-intended, but at times contradictory (and without doubt confusing) bureaucracy (Wyatt, 2008). For example:

- It is estimated that in 2007, for the first time in history, over 50 per cent of the world's population is living in urban areas.
- It has also been estimated that in 2005, over 60 per cent of carbon emissions came from industrial, commercial or residential buildings or operations.
- An estimated 25 per cent of these carbon emissions are caused by the way people heat, light and run their homes.
- In 2006, the UK was ranked sixth in the world league table of carbon emitting countries.
- The UK government's aim is to achieve a 60 per cent reduction in carbon emissions by 2050, compared to 1990 levels.
- Its target is for all new homes to be zero-carbon by 2016.

At the same time, it should be noted that new housing accounts for only one per cent of the housing stock in any one year, suggesting that the government's overwhelming focus on changing the approach to building new housing is mistaken: the role that refurbishment of the existing building stock should play in the reduction of carbon emissions is under-recognised.

Nevertheless, with the growing awareness of the role of the built environment in climate change, and an equally fast-growing body of understanding regarding the steps that can be taken to mediate its impact, a number of significant legislative and industry-body guidelines have emerged.

Importantly, the target for carbon emissions will become legally binding under the UK's Climate Change Bill (House of Lords, 2007), which was first introduced to parliament in March 2007 and which was amended, after a period of consultation, in November 2007. The bill includes proposed cuts of 30 per cent in carbon emissions by 2020.

Also, under current EU climate change proposals, it is likely that the UK will be required to increase the proportion of energy supplied from renewable sources to 40 per cent by 2020 (Greenpeace UK, 2008).

Both the carbon reduction and renewable energy targets will have a significant impact on the planning of, and manner of building of, new housing, including (for example) a drive towards combined heat and power (CHP) or district heat and power (DHP) generation, greater density of housing, and new types of materials used in construction. The combined cost impact of these changes is as yet unclear, although arguably, in the short term at least, costs are likely to increase due to a shortage of available skills, technologies and materials, whatever the long-term prospects might be for the industry to adapt to these new conditions.

Although pre-dating the Climate Change Bill and the latest EU legislative proposals, the government's Code for Sustainable Homes, published in December 2006, is intended to increase the environmental sustainability of homes. It also signals the future direction of building regulations in relation to carbon emissions from energy use in homes. From April 2007, the code, which has a scoring system of 6 levels, replaced the BRE's EcoHomes rating. Under the code, a home can achieve a sustainability rating of from one to six stars. The July 2007 Housing Green Paper announced a consultative process regarding making the Code for Sustainable Homes mandatory. The code applies to England only.

In September 2007, the National Housing Federation estimated that only two per cent of privately built new homes meet the new minimum criteria (NHF, 2007).

Although data are currently scarce, English Partnerships and the Housing Corporation (HC and EP, 2007) has estimated the cost of achieving Code Level 3 versus the EcoHomes 'Very Good' level as an increase of between 0.4 per cent and 5.7 per cent, depending on the type of house and the degree to which energy-efficiency measures have been applied. The sharing of energy services, such as combined heat and power (CHP) was generally shown to be the most cost-effective way of achieving compliance with Code Level 3.

More controversially, it has been estimated that increased green regulation could add up to 60 per cent to the construction cost of buildings (Miller Developments, October 2007). It is worth noting, however, that construction costs are not a major factor in the development of large-scale schemes. Land cost is the major part and, if the government had the appetite, it could engage in a degree of compulsory purchase, releasing land to the private sector and short-circuiting the existing planning constraints, on the condition that a number of mixed-use green enterprise zones were created. This radical activity is unlikely to be attractive to voters in the affected areas and it would take a brave (or highly principled) government to pursue this rapid solution to the UK's housing shortage.

A system of Energy Performance Certificates is also being introduced under Article 7 of the EU's Energy Performance in Buildings Directive. Certificates grade performance on a scale from A-G similar to the system used for grading white goods. Two types of Energy Performance Certificate (EPC) will be required for commercial buildings: asset certificates and operating certificates. The asset certificates will measure the intrinsic performance of the building based on its design, and the operating certificate will measure how the building is managed and is actually performing. The introduction of EPCs for commercial buildings will be implemented on a phased basis from 6 April 2008. In the case of domestic buildings, it was originally proposed that asset certificates would be required from June 2007, as part of the introduction of Home Information Packs. Implementation was, however, delayed until 14 December 2007, after which date all homes bought and sold in England and Wales now require an EPC. These domestic asset certificates include information relating to current average costs for heating, hot water and lighting.

The delay in the domestic introduction of EPCs was largely due to a lack of availability of trained assessors and to a lack of clarity regarding the role and content of the Home Information Packs. Perhaps this was an indication of good intentions outrunning available resources and a salutary lesson for future regulatory systems.

So far, this chapter has dealt 'top down' with an overview of the primary drivers for and mechanisms by which the greening of the housing market is to be achieved. To what extent are consumers aware of and reacting to this vital change? Is there really 'green' housing demand?

Consumer preferences – trends and changes

It is often assumed that the building of new homes is the primary factor in the drive for a greener and lower carbon housing market. Consumer demand is, however, not purely confined to new build. It is worthwhile considering whether there is any evidence of demand for a greener homes market in the rental sector and whether there is any regional variation in 'green' demand, before moving on to a fuller review of the available evidence regarding green demand for new homes.

The rental sector

Over the last ten years, the long-term upward trend in real house prices in the UK, driven by a shortage of supply, has created problems of affordability. This, combined with the decline in social housing provision and the availability of low-cost mortgages on non-primary residences, has created a highly fragmented and thriving private rental sector. The few studies publicly available of rental stock availability and pricing have not revealed any evidence of a green element in terms of either the purchase of properties for rental or the choice of properties to be rented. The government's announcement of a major review of the rental sector (January 2008) is expected to result in recommendations leading to the provision of affordable homes in substantial numbers. Since this provision would, it is anticipated, be modelled on the thriving professional rental sector in mainland Europe, it would involve pension and other investment funds in the building and management of homes. With corporate reputation at stake, therefore, it is likely the homes to be rented would be built to higher standards than would otherwise be the case. Yet again, therefore, it is anticipated this will result in a 'top down' rather than 'bottom up' push towards lower carbon residential developments.

Not a uniform national market – more a patchwork of regional demand

In green housing, as in other ways, there is no one national UK housing market. Different socio-economic demographic factors – for example, population, age profiles and net migration flows – affect local markets, producing a varied patchwork of local and regional demand. Surveys of housing needs, and the housing strategies which are published by Regional Development Agencies, detail the localised nature of demand and local aspirations, but there is – perhaps as yet – little or no reference in the majority of these documents to the need for an approach towards the greening of the housing market. No national survey has been conducted into the regional nature of green demand and no centralised data are available. A general mapping of current green developments shows a high concentration in the South East of England, but once again there is no study which reviews whether this is purely driven by greater population concentration, higher disposable incomes or any other socio-economic factor.

Green demand

Given that green demand has, so far, been very limited, the greening of the housing market has largely been led by government legislation and policy. For example, the introduction of a new relief from stamp duty land tax in the March 2007 Budget was specifically designed to help kick start the market for zero-carbon homes, to encourage micro-management technologies and to raise public awareness of the benefits of living in zero-carbon properties. The stamp duty land tax relief is time-limited to five years and provides exemption from tax liability when a house costs less than £500,000, together with a £15,000 reduction in tax liability to all qualifying homes worth more than £500,000. The relief is only available at the first point of sale on new homes. The certification process necessary to designate a home 'zero-carbon' is related to the assignment of the property's Energy Performance Certificate.

Since EPCs and Home Information Packs contain information which can lead to energy cost savings, it is likely that consumer reaction will be driven more by financial pragmatism, rather than a belief in the benefits of a lower carbon lifestyle. It is perhaps a hope rather than a documented trend that increased public awareness of the impact of carbon emissions from the built environment on climate change will encourage consumer-led demand for both highly sustainable new homes and for the upgrading of the existing housing stock.

The limited evidence available of changing consumer preferences in the UK is largely anecdotal. For example, Debbie Aplin, Managing Director for Regeneration at Crest Nicholson, has been quoted as saying:

> *'People can be resistant to change, they are worried that technology may be too complicated, that it will break down or that it won't deliver what has been promised. However, we have found that, over the past couple of years, people are taking environmental matters seriously enough to influence their buying decisions. There is a discernible change in the way that people, particularly the young, want to live their lives'* (Fellows, 2007).

Some recent research supports her view. For example, in a survey by the Sponge Sustainability Network, 54 per cent of respondents claimed they had not considered the importance of energy and water-saving features when they chose their existing homes. If they were to purchase now, however, only 25 per cent would still view energy efficiency and water consumption as unimportant.

In Europe, perhaps the most highly documented sustainable community is that at Hammarby Sjostad in Sweden. A low-rise water-front apartment-style development, built in a variety of architectural styles and from a variety of materials, Hammarby incorporates energy-saving and water-saving technologies, including the provision of central heat and power along with a system which uses combustible waste as input to the combined heat and power plant. Apartments at Hammarby have proved highly popular with sales values marginally above comparable local developments. There is no study, however, which isolates the other purchase decision factors such as location and ease of transport access from Hammarby's green credentials.

In summary therefore, there is as yet little evidence of a UK consumer-led demand for green housing. Purchase or rental decision-making seems still to be based on cost and locational factors.

The sale of eco-homes

Given the shortage of documented demand, how then have sales of green homes performed in the market?

In the UK, only one scheme – BedZed, at Wallington in Surrey – has been in existence long enough and on a large enough scale, to provide a basis for review. BedZed (the

Beddington Zero Energy Development) developed by the Peabody Trust, was completed in summer 2002 and is the UK's first carbon-neutral community. A mixed-use, mixed-tenure development, BedZed's design concept was driven by a desire to create a '...*zero fossil energy development*'. BedZed comprises 82 residential homes with a mixture of tenures, plus a further 14 galleried apartments for outright sale. The buildings are constructed from thermally massive materials that store heat during warm conditions and release heat at cooler times. The houses are arranged in south-facing terraces to maximise heat gain from the sun and each terrace is backed by north-facing offices, where minimal solar gain reduces the tendency to overheat and the need for (energy hungry) air-conditioning. In addition, a number of measures have been introduced to reduce energy, water and car use. A CHP plant provides energy for the development.

Both Peabody and its consultants are monitoring the performance of BedZed. For example, property consultants Savills (previously FPD Savills) have compiled a marketing report, the main findings of which are:

- Over the two-year period from 2002 to 2004, prices at BedZed generally increased in line with the market average and at a premium of three per cent to the local market average, when compared with a similar standard of conventional housing.
- Two-bedroom apartments at BedZed appear to have underperformed in the market, with realised prices increasing less than those of conventional housing types. This is thought to be related to the type of buyer usually attracted to the two-bedroom apartment market. Typically, buyers of such apartments are young, single people on relatively low incomes who are perhaps not attracted by the type of community BedZed represents, and might prefer a more central and vibrant location. Typically these units attract a premium of about 8.5 per cent.
- In contrast, three- and four-bedroom apartments at BedZed have out performed the market. Again, this is thought to be related to the buyer profile for this sort of space, which appeals to young professionals who are married with a young family. As well as the value systems associated with the purchase of an apartment at BedZed, the generous space standards provided by the scheme's units are believed to be an important contributing factor to the values achieved. Three- and four-bedroom apartments at BedZed achieved premiums of about 18 per cent and 17 per cent respectively.
- One-bedroom apartments achieve the highest premium – 20 per cent – which is due in part to their larger than average size.
- The overall finding is that the premiums realised are derived from three main elements – the target market, location and the features of the scheme, including space standards.

Box 14.1: BedZed – sources of premium pricing

- Typical buyers are professionals in full-time employment, some with young families, with ages ranging between their late 20s and mid-50s.
- Only one-third of occupiers were attracted to the scheme by its environmentally friendly qualities.
- Part of the attraction of BedZed is its proximity to public transport and access to central London. Without this access to public transport and to areas of mass employment, premiums would be lower.
- Value-adding features at BedZed include the provision of gardens for smaller units in the conventional market, which are historically believed to have created a tangible increase of between four and five per cent. Conversely, the provision of gardens with the larger units is not seen to add value to the premium, since gardens are commonly available with most similar sized units in the area. Sun spaces at BedZed are considered an attraction by potential buyers and are thought to facilitate sales. A related tangible increase in property values has not been proven, however.

With a number of low- or zero-carbon developments at a marketing stage, it will be interesting to track realised prices for these alongside those for standard developments.

Marketing low-carbon homes

Just as there is little material available regarding the pricing and sale of low-carbon homes, there is little historic material available regarding the marketing of low-carbon homes. Currently, low-carbon developments with sufficient scale and at the appropriate stage of development to engage in general marketing include One Brighton, and Middlehaven (RiversideOne), with Crest Nicholson's Bath Western Riverside in the early stages of defining its marketing strategy. One Brighton is a joint venture between Crest Nicholson and BioRegional Quintain, RiversideOne is a BioRegional Quintain project. None of these projects is as yet at a stage where lessons can be drawn from the relative success or failure of the various approaches chosen.

The marketing for each project nevertheless provides an interesting contrast and reinforces the view that residential property marketing is largely driven by the localised nature of the market.

With the buildings currently under construction and scheduled for completion by August 2009, One Brighton is the furthest advanced of the schemes. The development comprises 172 apartments and 24,000 square feet of commercial space on a plot of land near Brighton's railway station. Active selling of the apartments is expected to begin by the end of 2007. The website for the development can be viewed at www.onebrighton.co.uk and, as is immediately obvious, the development is being sold largely on the application of the ten guiding principles of One Planet Living® (see Box 14.2). Helen Devy, Head of Residential at Quintain Estates and Development PLC has been the primary architect of the scheme's marketing. She believes that the location of the site, combined with a local shortage of mid-price, high-quality housing of this type in the area, and local demographics, including the presence of a large number of urban professionals associated with the various local universities and colleges, for whom an environmentally friendly lifestyle is important, plus a number of 'green' commuters, mean there is a sufficiently large target market for this approach to succeed. Marketing materials position One Brighton as a sustainable One Planet Living® community in Brighton's New England Quarter and feature, for example, the integral roof-top allotments and the sense of community the building design is intended to generate.

A number of the apartments at One Brighton will be sold 'off-plan' and a significant number of enquiries have so far been received, giving an early indication of the likely success of the marketing programme.

Box 14.2: The ten One Planet Living® principles:

- zero-carbon
- zero waste
- sustainable transport
- local and sustainable food
- sustainable water

- natural habitats and wildlife
- culture and heritage
- equity and fair trade
- health and happiness
- local and sustainable materials

Plate 14.1: The planned 'One Brighton' development

Plate 14.2: Riverside One at Middlehaven

In contrast to the 'green' marketing emphasis at One Brighton, the approach chosen for RiversideOne at Middlehaven is more traditional. The Middlehaven development is located in a regeneration area (see case study at the end of the chapter). There is no local housing shortage per se, however, the bulk of the mid-priced housing stock is of a traditional Northern terraced type which is perceived as less attractive by the younger professionals who form the development's target market. Based on extensive market research, the primary marketing messages for RiversideOne are centred on the scheme's contemporary architecture and the waterside living experience it will provide. 'Collective individuality' is emphasised, with the scheme's sustainability and zero-carbon nature providing the secondary marketing message, despite the fact that RiversideOne currently represents the UK's largest zero-carbon development. The first apartments from this scheme, which will be in the 'Community in a Cube' building, are scheduled for release to the market in January 2008. The RiversideOne website can be viewed at www.riverside-one.com

Bath Western Riverside, which is located on a brownfield site to the west of Bath's city centre, will comprise between 2,000 and 3,000 apartments and homes, including 300 affordable homes, a primary school, shops and a new public riverside park, forming a new community. Crest Nicholson will deliver the residential development in the western quarter of the site. Through the planning and design of the development, Crest Nicholson are endeavouring to ensure residents can choose a sustainable lifestyle with features including energy efficiency, rain water harvesting, waste recycling and a sustainable living centre. Education and support are perceived by Crest Nicholson as key to the marketing of sustainable property and at the Bath site it is planned to include a manned centre where staff will be available to educate home owners about sustainability and give them the support they will need to ensure the site remains sustainable in the long term. With the first phase of the development due for completion in 2011 and the final phase in 2021, Bath Western Riverside should provide an interesting long-term study in the marketing of sustainable homes and the changes of approach that will inevitably occur over time.

The One Gallions scheme in east London's Docklands was intended by Mayor Ken Livingston to set the standard for environmentally sustainable development across London. The scheme, which is being developed by Crest Nicholson, BioRegional Quintain and Southern Housing, will comprise over 200 apartments and will incorporate a number of features which will help residents to lead an environmentally friendly lifestyle, including communal roof-top greenhouses where residents can grow their own fruit and vegetables, an energy centre, integrated waste management systems, a car club and a One Planet Living community centre. Marketing has not yet begun for this scheme, which is scheduled for completion in early 2010. Crest Nicholson believe, however, there is an increasing consumer demand for sustainable development and that *'…people are beginning to understand that climate change needs to be addressed'*.

Plate 14.3: The One Gallions scheme

Other schemes already on the market with a high sustainable element include developer First Base's 147-unit, Adelaide Wharf at Shoreditch, London. This scheme is over 50 per cent affordable, including a significant proportion of units dedicated to key workers. Green features include recycling facilities, 'green' living roofs and various energy-efficiency measures. The private element of the scheme has, however, been marketed on its location and the opportunities afforded by the overall regeneration of Shoreditch, rather than as a 'green' living experience.

With so little evidence available of the success or failure of marketing approaches to the sale of green homes, it is perhaps relevant to consider the parallels that can be drawn with the organic food movement.

For many years the organic movement was seen as a niche market, yet as concerns have grown regarding the impact of additives and chemical fertilisers on human health, animal welfare and the protection of the environment, so has demand for organic produce. In other words, market growth has been linked to perceived personal benefit. Cost factors have, however, limited the market, with a 2006 Mintel study showing older and more affluent consumers (in the 55 to 64-year old age group) buying organic produce more frequently than younger, less affluent purchasers (Mintel Reports, 2007). Despite an estimated 30 per cent growth in demand in 2005-2006, organic produce was estimated to account for just over one per cent of overall food and drink sales.

Little of this increase has been attributed to direct marketing, with general education and awareness seen as leading demand.

What then are the lessons for the marketing of green homes? It is arguable that in the short term, the higher cost of highly rated sustainable homes will make them the preserve of the older, more aware and more affluent consumer. General mass demand will be linked to a lowering of costs and to a greater understanding of the personal benefits to be derived from a green lifestyle.

Green mortgages

Even if there is little current evidence of consumer demand for green housing, it is interesting to consider whether consumers might be expressing their environmental concerns through their financial decision-making, including the selection of so-called 'green mortgages'. There is no clear definition of the term 'green mortgage'. Also known as ethical or sustainable mortgages, the type of offer varies by provider. Whereas some of the 'green' mortgages offered are purely available for financing the purchase of properties that are respectful of the environment, others are constructed so that for each mortgage lent there is an offsetting activity such as the planting of trees.

In the USA, states such as Pennsylvania, New York and Kansas and commercial banking organisations such as Citigroup, Bank of America and JP Morgan Chase offer 'green mortgages'. These usually take the form of a lower-cost or subsidised loan to improve the environmental-friendliness of a property, most usually its energy efficiency. Citigroup and Bank of America both offer a discount on closing costs or fees for environmental features. Some commentators have remarked that the added benefit to home occupiers of lower energy bills is attracting certain mortgage providers, who are vulnerable to the subprime mortgage market, in order to promote 'green mortgages' to lower income households, since the lower bills will give the occupiers more available income to meet their repayments.

In the UK the Council of Mortgage Lenders estimates the number of green mortgages as '*...a few thousand only*' and a tiny fraction of the UK mortgage market. Green mortgages are generally more expensive than other traditional mortgage offers and consumer decision-making in this market is primarily cost-based, with few consumers as yet placing their environmental priorities above economic necessity.

In his last Budget as chancellor, Gordon Brown claimed he would encourage lenders to create '*...a new market – mortgages for immediate capital investment in energy efficiency that cuts consumption and cuts bills*'. At present, however, only the three previously mentioned lenders – the Norwich and Peterborough Building Society, the Co-operative Bank and the Ecology Building Society – offer green mortgages. With the exception of the Ecology, most green mortgages are based on a form of carbon offsetting, where trees are planted to compensate for the carbon emissions produced by the homes on which the mortgage is granted. The Ecology only provides mortgages to those whose plans will be of benefit to the environment. Borrowers receive a discount on mortgage funds used to equip houses with energy-saving measures.

A number of insurance companies are currently offering life assurance and other products through ethical funds that define where the money is invested, and which type of company activities are supported or avoided. There is at present, however, little choice for those who wish to insure their homes and contents through an environmentally friendly provider. Naturesave Policies Ltd, for example, pledges that ten per cent of the premiums generated are used to benefit environmental and conservation organisations with specific environmental projects.

There is no specific regulatory mechanism in place to compare or control green mortgage or insurance offers and no independent study exists which documents the effectiveness of their carbon offsetting activities.

With no financial incentives offered, the decision whether or not to obtain a green mortgage is an entirely individual and ethical one. Arguably, if we are to fully tackle climate change and the role of housing in mitigating this, there should be a fully integrated financing system for housing which encourages ecologically sound building and operating practises. The government should be encouraged to deliver on Gordon Brown's promises.

The Callcutt Review and the greening of the market

As noted earlier, the UK housing market has generally been characterised by a shortage of supply and rising prices. The Callcutt Review, which was announced in December 2006 by the Secretary of State for Communities and Local Government, was charged with:

- An examination of how the supply of new homes is influenced by the nature and structure of the housebuilding industry, its business models and its supply chain, including land materials and skills.
- A consideration of how these factors influence the delivery of new homes to achieve the government's target (200,000 new homes per annum), meeting house buyers' requirements and aspirations, achieving high standards of energy-efficiency and sustainability as set out in the Code for Sustainable Homes, and progressing to a zero-carbon standard.
- Making recommendations.

The review team published a call for evidence and over 80 responses were received in total from a wide range of organisations including individual housebuilding companies, developers, national trade organisations, national public sector organisations, professional institutions, architectural firms and trades unions.

A survey of available published responses to the review notably reveals a call for a restructuring of the planning regime to ensure there are sufficient sites with implementable residential planning permission, in the right locations, of the right sizes to fit local market demand. A comment from the Home Builders Federation notes that in their opinion the government's zero-carbon targets and broader sustainability goals can only be achieved if the national timetable, agreed between the industry and government, is adhered to, if local planning authorities are restrained from setting a multitude of independent policies and if home buyer interests are fully protected.

In his final report, published in November 2007, Callcutt states that England's housebuilding industry is capable of delivering 240,000 homes a year by 2016. Land is perceived as the key to delivery and that Planning Policy Statement 3, which was published in late 2006, has provided a sound policy framework for ensuring that an adequate supply of development land is available. Callcutt supports the use of brownfield sites as the key to providing a more environmentally sustainable lifestyle, by cutting down on commuting journeys and providing better access to services and leisure activities than new settlements and edge-of-town developments.

Callcutt also states that with multiple technical options and long lead times in the production supply chain, the (housebuilding) industry will be stretched to meet the government's aspiration to be world class in the delivery of zero-carbon homes by 2016. Callcutt calls for clarity in relation to zero-carbon rules and proposes that government should set out by 2008 exactly how zero-carbon performance is defined, along with a recommendation for the setting up of a new and independent delivery body which would monitor and guide the zero-carbon programme.

Generally, reaction to the Callcutt Review has been positive, with for example The British Property Federation welcoming Callcutt's call for flexibility and for developers to be able to take the lead on the best way to meet zero-carbon targets. Industry perception is that the observations of the Callcutt Review regarding the lack of skills and shortage of available technologies for delivery are accurate and that more time and investment is required to meet the government's targets. There is a perception, for example, that Level 6 of the Code for Sustainable Homes is currently unachievable with existing materials and technologies, and that if achieved, due to the low levels of water consumption in particular, the resulting housing will not be acceptable to the consumer. There is also confusion within the industry regarding the definition of zero-carbon performance and clarification would be welcomed.

Conclusion

This chapter has examined some of the trends and initiatives relating to the 'greening of the housing market'. Driven primarily by government awareness of the impact of the built environment, and housing in particular, on carbon dioxide and other greenhouse gas emissions, and as a result by national and local government initiatives, the green market is in its early stages. Public awareness is still emerging and demand for 'green housing' is in its infancy. Hard data regarding the success, or failure, of a 'green housing market' do not yet exist. Several large-scale developments are currently underway and they will be closely monitored for lessons that can be learned to ensure the future housing stock is both environmentally friendly and popular with the public.

There are, however, certain issues still to be resolved. Many government proposals for 'green housing' are located in areas not previously designated for housing use. Arguably, because of the infrastructure demands which will be created by these developments, we should look again at urban density. It is often stated that if London (current population around 7,640,000) were built to the same density as Paris, it could house a population of around 30 million inhabitants! The issue is not just the cost of infrastructure provision; it is also the nature of the society which the proposed (relatively) low-density housing may produce. In general, however, the failure of the efforts of the 1960s would appear to have been forgotten. Those same efforts engendered large-scale housing estates far from city centres with little integration of schools, healthcare services, shops and jobs, having little or no public realm and often being poorly served by public transport. These factors created 'sink estates' and associated urban deprivation. The findings of the Callcutt Review, including its support for further brownfield developments, are very welcome in this respect. The developments cited in this chapter have sought to learn from the lessons of the past and should produce the exemplar schemes of tomorrow.

There is no doubt however that, despite current mass demand, and driven by the imperative of climate change, whether urban or sub-urban in nature, the future housing stock will be – Green!

Case study – Middlehaven

Middlehaven represents one of the most ambitious sustainable mixed-use regeneration projects to date. The scheme is located on a 40-acre site which is centred on the redundant former Middlesbrough docks. The site has been derelict since the decline of the steel industry in the area in the late 1970s and 80s. It is located near to the site of the 35,000-capacity Riverside Stadium, completed in 1995 and home to Middlesbrough Football Club.

The regeneration programme is designed to attract inward investment to the area, generate employment and stem the tide of outward migration, particularly of young professionals. In a particularly competitive inward investment market, overseas investment is seen as a key target for this scheme. The proximity to the football stadium is considered to be particularly critical, since television coverage of football games will include background shots of the area and thus provide an opportunity to showcase the development.

The site was purchased by English Partnerships, who invested in the decontamination of the site and who commissioned an original masterplan by the leading architect, Will Alsop. Tees Valley Regeneration, which is leading the Middlehaven regeneration project, is one of the government-sponsored Urban Regeneration Companies. Its role is to spearhead major physical regeneration and bring inward investment to the Tees Valley. The company is owned by English Partnerships, One NorthEast and the five local authorities who make up the Tees Valley – Darlington, Hartlepool, Middlesbrough, Stockton and Redcar and Cleveland.

In November 2006, BioRegional Quintain Ltd were appointed as lead developers for the £200 million development scheme, for which Treasury grant funding is in place.

Middlehaven will be the largest zero-carbon development in the UK and the total scheme is expected to deliver more than 1,000 new jobs, approximately 750 new homes, in excess of 200,000 square feet of new office and leisure space, including bars, restaurants and a hotel, and 25,000 square feet of shops.

→

Detailed plans were submitted in May 2007 for the first two buildings on the site, which comprise 150 residential units and 14,000 square feet of commercial space. Heads of terms have also been signed with Hilton Hotels for a 150-bed Garden Inn hotel. Work began on site in late autumn 2007.

The development is emphasising iconic design as well as sustainability, with a strong emphasis on leading-edge design seen as a significant differentiating characteristic for Middlehaven. Pete Halsall, Chief Executive of BioRegional Quintain, has said, *'There will be no other place in Europe with so many daring and exciting, yet sustainable and practical buildings, standing side-by-side'*. Architects currently working on the development, in addition to Will Alsop, include Studio Egret West and Fielden Clegg Bradley. Jestico and Whiles have been named as architects for the hotel cube building.

The design of the site is intended to introduce a new element to residential development in the area – a high-quality riverside living experience. English Parnerships and Tees Valley Regeneration selected BioRegional Quintain as their lead developers after realising that the building of a zero-carbon development, as well as the iconic architecture, would additionally differentiate Middlehaven from other European schemes. Middlehaven won the 'big urban projects' category at the MIPIM (Architectural Review) Future Projects Awards 2007 in Cannes, France, in competition with schemes from across Europe, and has generated a huge amount of publicity worldwide since its inception.

As the UK's largest zero-carbon development, Middlehaven will utilise all of the One Planet Living® principles.

Middlehaven's green credentials

- Zero waste
  - Best practice targets for reduced construction waste.
  - Internal and external storage bins for segregating different municipal waste streams.
  - On-site composting proposed, subject to further licensing and viability study.

- Sustainable transport
  - Car club.
  - Facilities to support home-working.
  - Public transport connections.
  - Cycle storage.
  - Cycle club.

- Sustainable water
  - Water-efficient fittings and appliances used throughout.
  - Surface and rain water managed sustainably.

→

- Equity and fair trade
  - Contractors used and sought who invest in staff development.
  - High levels of accessibility.
  - Fair trade food and goods promoted in restaurants and shops.

- Local and sustainable materials
  - Stretching targets for use of recycled and reclaimed locally-sourced materials.
  - Embodied energy target.
  - Certified sustainable timber.
  - Promotion of local suppliers and contractors.

- Natural habitats and wildlife
  - Strategy developed with local ecologists and local stakeholders.
  - Habitat creation.
  - Integrated nest boxes.
  - Drought resistant and native planting.

- Health and happiness
  - Community trust established.
  - 'Natural' materials used.
  - High levels of indoor air quality, daylight etc.
  - Buildings and public realm promote safe, secure and healthy environments.

- Culture and heritage
  - Community involvement as part of design development.
  - Community extranet established.
  - Events programme developed.

- Local and sustainable food
  - Landscaping and public realm to include opportunities for food-growing.
  - Balconies to include herb boxes.
  - Facilities to receive local food box deliveries.
  - Promotion of local food market in events space.

- Zero-carbon
  - High performance building fabric.
  - Energy-efficient lights, fittings and appliances.
  - Localised renewable energy centres (creating all the energy to heat, cool and power the development from a combination of on-site and off-site renewable sources).
  - Biomass boilers/fuel cells to generate energy to meet space heating and hot water demands.
  - Aim to establish local biomass/other renewable fuels supply chain.
  - Energy Services Company (ESCO) established to ensure ongoing sustainable energy provision.
  - 'Green' electricity bulk purchased via ESCO from net new renewable capacity.

References

Barker, K. (2004) *Delivering stability: securing our future housing needs*, HM Treasury. Available from: http://www.hm-treasury.gov.uk/consultations_and_legislation/barker/consult_barker_index.cfm

BPF (2007) *British Property Federation submission on the Calcutt Review of Housebuilding*, British Property Foundation

CABE (2008) *Hammarby Sjöstad, Stockholm*, Commission for Architecture and the Built Environment. Available from: http://www.cabe.org.uk/default.aspx?content itemid=1318&aspectid=7

Calcutt, J. (2007) *The Review of Housebuilding Delivery - the Callcutt Review*, Communities and Local Government. Available from: http://www.callcuttreview. co.uk/default.jsp

CLG (2006) *Code for Sustainable Homes: A step-change in sustainable home building practice*, Communities and Local Government. Available from: http://www.planningportal.gov.uk/uploads/code_for_sust_homes.pdf

CLG (2006) *Planning Policy Statement 3: Housing*, Communities and Local Government. Available from: http://www.communities.gov.uk/publications/planningandbuilding/pps3housing

CLG (2007) *Homes for the future: more affordable, more sustainable – Housing Green Paper*, Communities and Local Government. Available from: http://www.communities.gov.uk/publications/housing/homesforfuture

CLG (2008) *Private rented sector to be focus of independent review*, Communities and Local Government. Available from: http://www.communities.gov.uk/news/corporate/670940

EC (European Commission) (2003) 'Council Directive 2002/91/EC of 16 December 2002 on the energy performance of buildings', *Official Journal of the European Communities,* No L 1, 04/01/2003, pp. 65-71

Fellows, S. (2007) 'Des Res', *Property Week Insight Series – Guide to Green Development*

FDP Savills (2003) *ZED Units: A Market Analogy*, FDP Savills

Greenpeace UK (2008) *Renewable energy target 'entirely achievable'*. Available from: http://www.greenpeace.org.uk/media/press-releases/

HBF (2007) *Calcutt Review of Housebuilding Delivery – Submission by Home Builders Federation*, Home Builders Federation

HC and EP (2007) *A Cost Review of the Code for Sustainable Homes*, Housing Corporation, English Partnerships

HM Government (2007) *Draft Climate Change Bill*, TSO. Available from: http://www.official-documents.gov.uk/document/cm70/7040/7040.pdf

HM Treasury (2007) *Budget 2007 – Building Britain's long-term future: Prosperity and fairness for families*, London: HMSO

House of Lords (2007) *Climate Change Bill*, London: TSO

Mintel Reports (2007) *Organics – UK*, Mintel

NHF (2007) *Government 'in danger' of missing target for all new houses to be zero-carbon by 2016*, National Housing Federation. Available from: http://housing.org.uk

Peabody Trust (2002) *BedZed (Beddington Zero Energy Development) factsheet*. Available from: http://www.peabody.org.uk/pages/GetPage.aspx?id=179

Sponge (2007) *Eco Chic or Eco Geek: The Desirability of Sustainable Homes - Executive Summary*, Sponge Sustainability Network. Available from: http://www.spongenet.org/library/Sponge%20Eco%20Chic%20or%20or%20Eco%20Geek%20The%20Desirability%20of%20Sustainable%20Homes%20Executive%20Summary.pdf

Sweett, C. (2007) *A cost review of the code for sustainable homes: Report for English Partnerships and the Housing Corporation*, English Partnerships, Housing Corporation. Available from: http://www.cyrilsweett.com/pdfs/Code%20for%20sustainable%20homes%20cost%20analysis.pdf

Wyatt, A. (2008) 'We need a green industrial revolution, not more green tape', *Estate Review*, 7 April. Available from: http://www.estatesreview.com/news/140/ARTICLE/1931/2008-04-07.html

CHAPTER 15:
'Greening' your organisation

Caleb Klaces and Adam Broadway

Introduction

How can you tell a green organisation in the housing sector when you meet it? How can you become one? This chapter begins to answer these questions.

The most successful organisations will be those that think about everything they do in terms of sustainability principles. But we argue that it is key to focus on modest, carefully chosen, authentic steps and creating a culture, rather than a set of vague targets that feel impressive but can leave an organisation confused and ineffective.

With this in mind, our chapter begins with an overview of the strategic context for sustainability in the housing sector. It then gives a short guide to good processes and practical actions that an organisation can follow to respond to the challenge of operating in ways which meet our needs while allowing future generations to meet their own needs.[1]

We hope this will help organisations develop a strategic response to the market and to policy, in the context of sustainability and what we call here the 'environmental imperative' (the belief that there are limits to what the natural environment can withstand). We also hope that our snapshot of a rapidly changing market and the clear benefits of corporate sustainability in the first section is an example for organisations' own analyses using the tools we set out in the second. Much of the *why*, and indeed the *how* in response, will not change dramatically over time. But it would be naive to recommend the same checklist of actions for all organisations. What can, and we believe should, be applied right across the spectrum of organisations which build, sell or manage homes is a set of questions to ask, and rationales for deciding on and prioritising actions that comes from a sustainability perspective. So you can do what is appropriate for your organisation.

In the last few years, our knowledge of our effect on the environment has grown enormously (UNEP, 2007). In the same period, an unprecedented consensus has developed among scientists about the reality of climate change and the small window of opportunity we have to mitigate the worst impacts (IPCC, 2007).

---

1   This is an adaptation of one of the best known definitions of sustainable development, from the Brundtand Commission (World Commission on Environment and Development, 1987). It defined such development as that which '...*meets the needs of the present without compromising the ability of future generations to meet their own needs*'.

There are particular challenges which need to be tackled by the housing sector, including reducing the environmental impacts of existing stock, both in their fabric and by changing the behaviour of people who live in them, and building and maintaining new places that are well-integrated and encourage more sustainable lifestyles.

To create organisations that can truly engage with these agendas and reap the full benefits requires staff who are knowledgeable about the issues, and feel comfortable in an organisation that is itself making changes. A 'green strategy and action plan' and development of better systems can lead to a different corporate culture, where practices become second nature.[2]

'Greening' and housing organisations

While the Callcutt Review found that *'England's housebuilding industry is in shape to deliver the homes we need for future generations and is capable of delivering 240,000 homes a year by 2016'* (Callcutt, 2007), the current supply of new homes is some way off the government's targets, both in terms of supply (only 163,400 new homes were built in 2005/06 and 167,600 in 2006/07) (CLG, 2008a) and environmental standards (HC, *et al.*, 2007).

Government targets are for three million new homes by 2020 (CLG, 2007a). It is recognised that all the new homes built now could represent one-third of the housing stock by 2050 (CLG, 2006) and that improving the environmental standards of these new homes (with the aim of all new homes in England being zero-carbon by 2016) will contribute significantly to the government's carbon reduction targets and avoiding the release of at least 15 million tonnes CO_2 per year by 2050 (CLG, 2007a).

It is well-documented that that there is currently a significant under-supply of homes in the UK. Partly because of this shortage, house prices in 2007 reached record levels. The average UK house in 2006 cost £204,813 (CLG, 2008b), which means that a house cost 6.9 times the median income (CLG, 2007b). Where and how these new homes are built is opening up many old arguments about greenfield versus brownfield development and, just as importantly, the focus of housebuilding activity and regeneration priorities between the north and the south of the country.

The supply of most of the UK housing market is now in the hands of a few major developers. Over the last few years there have been a number of notable housebuilder mergers, such as Taylor Woodrow and George Wimpey to form Taylor Wimpey – the largest housebuilder in the UK, (Smith, 2007) – and more recently Barratt acquiring Wilson Bowden Plc.

---

2   This can sit either inside or alongside an organisation's corporate social responsibility (CSR) programmes. It becomes part of the economic, social and environmental 'triple bottom line' way of thinking about business.

At the same time, the housing association sector has also undergone significant consolidation with a number of notable mergers, including several of the larger associations. In addition, in response to the government's agenda to achieve more affordable homes for less public money, a number of development partnerships have been formed.[3] In 2007, the Housing Corporation opened up the ability to obtain a social housing grant to a number of private developers, in an attempt to increase affordable housing supply (HC, 2007). While the number of players appears to be reducing, it remains a very competitive development world.

The emphasis on existing stock is growing, too, particularly at a regional level. London is one of the most advanced cities in the world in terms of policy and support for green action: from supporting people to make changes with online tools to the ambitious London climate change action plan (GLA, 2007). The 'transition towns' movement[4] is also significant, with a number of communities taking co-ordinated action to reduce their carbon emissions.

Still, in terms of policy, the sustainability agenda in its broader sense has to a large extent been hijacked by the carbon agenda, with the most forward thinking legislation and policy instruments being targeted at this area at the expense of addressing many broader social and indeed environmental issues.

At present the only real legislative pressure on the housing market to become 'greener' is through compliance with planning and building regulations. The *Code for Sustainable Homes* introduced in April 2007 is a step forward but is not mandatory except for organisations seeking public money. The *Code for Sustainable Homes* replaced the EcoHomes Rating system which, while a useful tool, has never really been adopted by the speculative housebuilders as a marketing weapon. It can be argued that in a system and market of under-supply, such tools add no real value, in fact many private developers argue strongly that they are an unnecessary cost and that the 'public' aren't interested.

The assumption of residents' indifference, while generally plausible, is far too simplistic. At worst, it is a wilful perpetuation of a vicious cycle in which suppliers of new and existing homes do not act to make them more sustainable because of a perceived lack of demand, and residents do not demand more sustainable places to live partly because they do not know how to differentiate between houses or know what is possible, or from a perception that it is futile to ask. In contrast, an Ipsos Mori poll in 2007 (consisting of a quantitative telephone survey of 501 home owners and qualitative research with four discussion groups) found that 72 per cent thought sustainable homes are *'good value for money'* rather than *'bad value for money'*, and that a majority would pay more for a more sustainable home. It also found that

---

3   For instance, the Northern Lights Group (NLG) is comprised of Arcon, Arawak Walton, Equity, St Vincent's, Muir, Johnnie Johnson and Accent (http://www.arcon.org.uk/aboutus/nlg.html)
4   For more info see: http://www.transitiontowns.org/

92 per cent wanted to see sustainability features offered as options on new homes and 64 per cent thought these should be compulsory (Sponge, 2007).

Clearly, there are always some gaps between what people say they want in polls and what they actually do. The powerful idea we are considering here is the potential for organisations in the housing sector to help along the shifts that are taking place, to benefit themselves, the environment, and people in general. This could create a new and virtuous cycle.

The business benefits

In July 2007, Beyond Green did a study with Consensus Research International which polled over 1,000 consumers and 500 business managers in the UK. We found that there was a clear discrepancy between how people are behaving at home and at work, with environmentally friendly habits much less common in the workplace. We also found that a majority of employees are keen for the organisations they work for to have sustainability policies in place and that where such policies exist, they act as a direct stimulant for good habits among employees (Beyond Green and Consensus, 2007). Our practical experience of greening organisations supports the view that there is an appetite, and even an expectation, from employees for sustainability in the workplace, that is not being met.

There are several ways to look at benefits. Some activities are responsive to pressure from customers and are therefore adjusting to the market and keeping up. Some are about avoiding future risks, some are the benefits of being proactive and shaping the market. A recent analysis by WWF-UK, the Housing Corporation, Insight Investment and Upstream Strategies of the top 20 home builders (by volume) summed up the market well when it said that '...while not enough companies are in a position to assess how sustainability adds value to their business, there are strong examples of clear benefits being felt' (HC et al., 2007). The examples it gives are clear, for instance Taylor Woodrow reducing the cost of waste per home it builds from £351 to £291 from 2005 to 2006, despite increasing landfill tax, or many housebuilders finding that sustainability, for example in community consultation, has been instrumental in achieving planning permission.

If that specific, but rather simplistic clarity is at one end of the scale, then broad, bold, but vague pronouncements from business leaders are at the other. On 31 January 2007, Lehman Brothers published a report on the effects of climate change on business (Llewellyn, 2007). Speaking of the need for business to respond to environmental sustainability concerns, the report's author, John Llewellyn, said '...firms that recognise the challenge early and respond imaginatively and constructively will create opportunities for themselves and thereby prosper. Others slower to realise what is going on or electing to ignore it will likely do markedly less well' (ibid.). Lee Daley, Chairman and CEO of Saatchi & Saatchi UK went further, saying '...*companies which do not live by a green protocol will be financially damaged because their customers will punish them [...] They will not survive*' (Grande, 2007).

The opportunities in Llewellyn's measured, but positive prognosis and threats in Daley's more severe warning are often closely linked and interdependent.

As we have already mentioned, employees now expect their employers to be thinking about sustainability. The next generation of workers will have greater expectations and a more discerning eye. A recent survey of 54,240 16-21 year olds applying to further education found that for almost half, environmental considerations are '*important*' or '*very important*' when deciding what organisation to work for (Forum for the Future, 2007). Greener companies can compete for the brightest people.

Encouraging those people to work 'imaginatively and constructively' around the challenges of sustainability precipitates innovation. Through innovation come new markets. In the case of genuinely sustainable products and services, these markets mitigate risks and provide robust new ways of generating income (the classic example is BP's investment in renewable technologies as oil stocks deplete). Emerging carbon markets benefit businesses in a different way, rewarding those that reduce emissions most. Gordon Brown has described the next phase of carbon trading as at '*the heart*' of the government's approach to tackling climate change (WWF, 2007). If the scheme is run properly, the benefits of greener practices can be multiplied. Businesses opening up new opportunities with innovative products and services can earn extra from the reductions in emissions they can achieve with them. Even in its present form, the voluntary UK Emissions Trading Scheme earned Land Securities £33,000 in 2006 through reductions in their emissions (IEMA, 2006).

An organisation which is reducing its contribution to climate change, taking on people with new expertise and challenging them to understand and make opportunities of new challenges, will also want to work with other organisations that are doing the same. In fact, the reputational risks of not doing so are becoming increasingly acute. Suppliers and contractors won't work with you if you don't meet exacting standards, just as 'customers will punish' you. Conversely, as BT has found, demanding high environmental standards of a supply chain (IEMA, 2008) can lead to an increased rate of successful bids.

Understanding what the benefits and risks of specific actions are, can help to define an approach. For example, it may be that reducing emissions feels like a proactive move at the moment, but it may be merely reducing the risk of being penalised by carbon markets in the future. A broader agenda about defining and encouraging wellbeing (which may include carbon reductions) may be ahead of the market and be a unique selling point in the future.

It is not only the actions themselves that bring benefits but the way in which they contribute to a 'license to operate'. Organisations earn their right to do certain things from the way in which they gain support from investors, government, customers and staff. Talking about wellbeing does not give anyone the right to do something equally bad, it is not a balancing act. But earning goodwill, or indeed losing it, can result from really engaging with a difficult but important agenda. A risk assessment framework should therefore cover all issues, from the reputational risk of inaction to likely future legislative pressures, energy security and natural disasters.

Asking the right questions

The process of greening an organisation might best be summed up as systematically and practically responding to a series of questions:

- Why are you trying to do something/anything?
- What do you need to change that you can?
- How best can you achieve that with the resources available?
- How will you know that you have done it?
- How will everyone else know that you have done it?
- How can you make sure you keep doing it?

These are the questions which underlie the rest of this chapter, and the process is illustrated in Figure 15.1.

Figure 15.1: The process of 'greening' an organisation

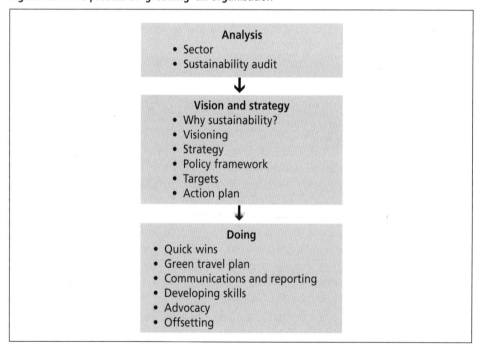

A vision: *why* into *how*

So, where do you start? A vision is a good place for several reasons.

When taken seriously, visions and vision statements are what you will become. If you don't know what success looks like, it becomes very difficult to achieve. Seeing it clearly makes it probable.

But before settling on a vision comes the question of *why*. As our discussion has tried to show so far, aspects of 'greening' are becoming less and less easy to ignore or overlook. It is the fashion. But that it is, does not on its own constitute a reason for change, nor a vision for action (unless your brand is based on being fashionable, whatever the weather). Aside from the most obvious reason of a feeling of duty, the impetus for 'greening' may be based on achieving any one of the benefits above, or avoiding the risks. But whatever it is, *why* should directly translate into *how* and *what*.

A good process would start by engaging with people at all levels in the organisation, and other stakeholders if appropriate. It would ask what they would like to be able to say in the future about where they work, in terms of sustainability. It would ask them what they thought they could realistically achieve in their area. You may find that what really matters and is most useful to your organisation is driving innovation through imposing increasingly strict efficiency targets on yourself (both in your operations and in what you deliver). In this case the corresponding vision may be as simple as *to use X% less natural resources year on year*. Somewhere else along the scale, an honest discussion with people in your organisation may make you realise that a powerful respect and love for the natural environment is the reason that most people, including the senior management, want to green the place, and a vision should reflect that. The vision might confront ecological issues of biodiversity, pollution and toxins. It may be that above being prepared for current and future legislation, the future defined by that vision is about using the stock that is under your control to increase the green space and habitat for wildlife. Under that vision actions may include literally 'greening' your office space, too, with air purifying plants.

The two primary purposes a vision should serve ask different things of it. It must be both inspiring and practical. In the best, the emotive power comes from articulating a common, challenging, but clearly achievable aim, while also indicating a focus and an approach. But it needn't try to cover everything that an organisation can possibly do – it should provide an impetus for change and slap cards on the table. You should set the scale of your ambition and then establish a vision of something that honestly reflects it.

Over the last 20 years, Interface, one of the world's largest carpet manufacturers, has been refining its vision for sustainability. Its most recent vision statement is

'...to become the first company that, by its deeds, shows the entire industrial world what sustainability is in all its dimensions: people, process, product, place and profits – by 2020 – and in doing so we will become restorative through the power of influence' (Anderson, 1998). This is an all-encompassing and profoundly ambitious vision, set on a timescale and supported by a strategy and policies.

But it would be inappropriate and meaningless for other organisations to adopt the same vision. Drum Housing Association won the Housing Corporation's Gold Award for 'Environmental Sustainability' in 2007.[5] The significant and innovative programme of work with its tenants and stock was achieved under a more sober but equally useful vision: *'Drum is committed to minimizing the negative impact and maximizing the positive impact of the Association's activities on the environment and biodiversity, and contributing positively to sustainable development'.* It goes on: *'We will do this by adopting an annual programme of prudent incremental activity designed to move the organization forward in support of our strategic objectives'.*

What Drum and Interface share is that they have visions appropriate to their ambition and what they can realistically achieve, and under which strategically chosen 'incrememental activity' – small, meaningful steps – fits.

What's your problem?

A vision grows out of the people in an organisation. But it must also be defined with a realistic understanding of what that organisation does and the effects on the world around it.

Having some data to work with is very useful for understanding priorities and targeting what are really the most environmentally damaging aspects of your work, and a good way of helping stakeholders understand the rationale for action. One way of doing that is through assessing your 'footprint'.

The youthful science of 'footprinting' aims to understand the extent to which specific human actions contribute to environmental degradation and climate change. An 'ecological footprint' is generally understood to mean a measure of the amount of productive land and sea area needed to generate the resources an activity or process consumes and to absorb the corresponding waste, usually expressed in global hectares (gha). Ecological footprinting techniques have calculated that if everyone in the world lived like a typical UK resident we would need three earths to sustain our consumption. A carbon footprint measures the total amount of carbon dioxide emissions caused by an activity or accumulated over

5   For more details see http://www.housingcorp.gov.uk/server/show/nav.3775

the life cycle of a product, and is expressed in kilograms ($kgCO_2$). The usefulness of both is in comparing the environmental impacts of different activities and processes.

When we worked with the Commission for Architecture and the Built Environment (CABE) on developing a climate change strategy (as consultants for Beyond Green), we advised that they footprint their office, travel and publications. Working alongside a process of stakeholder research to generate a public position and strategy for combating climate change, the footprint formed the basis of a plan to reduce the impact of their operations. What they found was that business travel, electricity, and their use of materials (paper, etc.) and waste were their largest impacts. But that also their publications came in a significant fourth. CABE could then focus attention with confidence, and create an action plan generated through discussion of the results with staff and also the manager of the building they rent.

In another project, GlaxoSmithKline UK used ecological footprinting as a way of understanding their products in order to define targets which they will work towards over the next three years. They found that packaging and ingredients had the highest footprints by a long way (with energy, waste and transport coming in under them). It followed that focusing their attention on recyclable packaging and sustainable sourcing strategies became the priority.

When it is done properly, a footprint is a powerful tool. But it is one that does not take into account the social and economic impacts of an activity, nor tells you without analysis the efficiency, and therefore potential improvements, of what you are doing. All oil extraction, for example, will have a massive ecological footprint, whether it is exemplary for what it gets out of the ore or not.

Some kind of audit is very useful, however, whether it is a footprint or a staff questionnaire, or conversations with your stakeholders. What should you think of as both part of your impact and within your control? Is the SUV the Chief Executive drives to work her business? These are interesting and productive questions that should be embraced. The key point is consistency. CABE wanted to engage with all its impacts, and so chose to find ways of reducing the flights that employees take. But the value of debate in workshops to work out how to understand and act on all aspects of the footprint goes further than that prioritising. We would argue that how people get to work is a valid concern of the organisation that employs them, but it may not be within the scope of your vision to try and reduce it. Discussing the question and taking its fruits seriously is part of the process of moving toward a culture of sustainability.

As CABE found, travel (both staff and residents') often causes an organisation's highest ecological and carbon impacts. In this area an audit of modes of travel, either through a questionnaire or monitoring over a period of time, is necessary. It is also important to find out *why* people are travelling as they are.

The plan and policy framework

Developing a 'green travel plan' is one example of a direct link between data from an audit and a policy and action plan. A green travel plan sets out ways in which an organisation will reduce carbon-intensive forms of transport, and change the methods of transport used by its staff and/or residents. For example, if many people are driving a short distance to work then it may be because there is no room for cycle storage. In this case, the action plan might be to carry out the necessary work, and the policy framework to set out the company's desire to reduce travel by car, and incentives for those that take up cycling. However, it might be that the policy that actually has most impact on the problem is about home working. Perhaps that is what needs to change, to allow people to work from home more often – rather than creating a new policy about cycling.

Where possible, plans and policies should be looked at holistically. There should not be more than is necessary, each measured against how they can best support self sustaining cultural change.

Action plans should set timescales, define responsibilities and potential areas which need new skills, methods of communication and engaging staff, and identify quick wins. Policies should support their implementation with meaningful targets and expectations.

An interesting example of an organisation developing policies and action plans at the same time was with a recent client of Beyond Green's. In order to write a top-level sustainability vision and strategy we held interviews with everyone in the company and brought together selected groups of people (many of whom would not normally work together) to discuss their aspirations for the company. Simultaneously, the company has been developing a technical guide to creating buildings to the highest government standards. Each part of the process informs the other, with the framework for delivery of core business working within a larger action plan generated from cross-discipline discussion. The streams are distinct but inform one another.

This model would work for housing organisations dealing with existing stock, too. Development of a focused response to any one of the challenges of upgrading stock (for example insulation, micro-generation, behaviour change, financing, adaptation) can and must be set in a wider context of strategy informed by the workings of the whole organisation.

Organising different streams like this needs careful planning. Creating policies provides a framework to all work. But they have to be the right policies. And they only half exist if they are just on paper. An environmental/sustainability/green manager should be appointed, with a committed senior member of staff (or board member) to report to and get support from, with champions at different levels of the organisation.

Doing

Achieving some quick wins will get tangible success and momentum. This might involve a guide for office staff about initiatives such as reducing water and energy use and recycling in the office. Or it may be offering a small new or adjusted service to residents. The importance is in a few short-term, high-benefit actions.

'Offsetting' emissions comes both into the quick win and final step categories. Offsetting is the act of mitigating greenhouse gas emissions by paying for emission reductions elsewhere, rather than reducing one's own emissions.[6] As it occurs now, we would argue that it should be thought of as a helpful tool in moving us to a low-carbon economy, but only after reducing direct emissions themselves as much as possible.

Apart from the conceptual issue of outsourcing an obligation, there is a problem with the timescales of offsetting. For example, reforestation projects will only soak up (a difficult to quantify amount of) carbon over a long period of time. A flight that is being offset, on the other hand, will emit the same amount of carbon all in one go, where it immediately begins to contribute to climate change. As a quick win, in the process of offsetting the emissions from your operations or stock you should be careful about choosing companies and projects which can show 'additionality': that they are only going to happen with the funding they receive from offsets. Renewable technologies and genuine energy-efficiency projects are thought to be the most effective at avoiding emissions comparable with those being offset. But such investment makes more sense as the last step in an implementation strategy rather than the first.

Offsetting is also a one-off act that does not necessarily contribute to lasting cultural change. The key to such change is developing resources that all employees can use and feel a part of: for example, encouraging staff to feel at home with the language of sustainability, through workshops, a glossary, and perhaps electronic resources.

Finding the balance of generic and applied, niche skills then becomes key. While it is true that at the moment taking a strategic approach and talking a coherent and consistent language of sustainability will give you some purchase in the marketplace, it is rare for organisations to have the range of skills needed to deliver sustainable communities (see Egan, 2004).

---

6   See the *Ethical Consumer Carbon Offset Guide* in May/June 2007 for an overview of providers.

Working with the supply chain at all levels, from building materials to office stationery, is often where green ambitions falter earliest. It involves asking questions about the materials, processes and social conditions under which the things you buy and use are produced. Well-established contacts and contracts, fears about having to make trade-offs with quality, lack of internal communications and support for the people involved, and of the tendency for bad habits to perpetuate themselves (which is not to be underestimated) – all play their part. But difficulties are more often than not perceived rather than real. Encouraging, and even demanding, different standards from your supply chain is an act of enlightened self-interest: the benefits to you will also ultimately be the source of benefits for your suppliers.

It can also provoke interesting debates. For example, a client recently had to choose between a water cooler which donated money to a charity supporting work relieving poverty in Africa, and having no water cooler for environmental reasons – to reduce plastic and energy use. The dichotomy is in large part a false one, as social and environmental concerns are inextricably linked (and ranking them is a fool's errand). But the choice says something about a company and can be made in light of its vision.

Being seen to do something

How a water cooler looks in your office is not as trivial a question as it may seem. Your physical work environment says something about the people and organisation working in it. The title of this section expresses the ambivalence of making your green ambitions, green actions and green failings public. 'Being seen' can be both negative and positive. The phrase comes from the intellectual and climate change campaigner George Monbiot (2006) who writes in his book *Heat* with typical straightness, '...what I am attempting to do is find the least painful means of making real cuts [in greenhouse gas emissions], rather than the least painful means of being seen to do something'.

As we have argued in this chapter, it is better to make real, if small, changes, rather than just be seen to do something. But we have also argued that there is no contradiction in considering how at each stage in this process those changes fit with, evolve and challenge the brand of your organisation, and fulfilling the expectations of your stakeholders. Indeed, that is much easier with steps in which you know what you are doing and why. The line between marketing and ethical communications becomes blurred, too. As communications specialist Dr. Ingrid Kajzer Mitchell has put it, '...marketing, then, becomes more than simply delivery of products or services, its value proposition includes ideas, opinions and ultimately behavioural change that simultaneously provide social and ecological value'.[7] It is our contention that no

---

7 Sustainable Futures, InterfaceFLOR, 2007.

housing sector organisation has yet claimed the centre ground through a truly green brand, nor delivered marketing that has simultaneously provided real social and ecological value.

However, reporting is becoming more and more important to shareholders. 70 per cent of the large home builders in the Building a Sustainable Future benchmark (HC *et al.*, 2007) reported publicly on their approach to sustainability (and 65 per cent have published policy). But the blindingly obvious principle that you should report things that mean something and which are true is not always followed. Reporting need not be flashy but contain the key quantitative as well as qualitative data on progress towards targets and implementation of a strategy.

Regular monitoring against any targets is crucial for the evolution of your ambitions, as well as making the targets mean something. In addition to internal analysis, there are a number of standards, awards and benchmarks that can be used to substantiate the quality of your approach, from local economic impact to sustainability reporting. ISO 14001 is a well-recognised place to start in verifying environmental corporate management systems.

There are a growing number of awards without obscure numbers in the title, too. Both *The Times* and *Observer* newspapers support high-profile awards for green companies. And right at the end of the scale there's the Queen's Award for Sustainable Development, which rewards a consistent and long-term commitment to performance and improvement.

And then what?

How can you make sure you stay green? There isn't an easy answer but we hope that the approach we have outlined helps generate dynamic systems. Sustainability is, after all, about practices that it is possible to perpetuate.

One way of staying ahead is like Drum Housing Association, which we return to. It has twelve Continuous Improvement Groups, one of which is concerned exclusively with sustainable development, and which include 60 residents. Such groups benefit both the association and its residents, who are willing to take part because they have seen real improvement in the places where they live.

Better practice increases expectations, which encourages better practice. If initial actions are well-targeted and successful, then it becomes easier and easier to convince everyone that such actions are a better idea. And of course, there is always more to be done.

Case study: The green route planner – Gentoo Group

Sally Hancox

The North East-based Gentoo Group was winner of the 2007 UK Housing Awards for Increasing Environmental Sustainability and the 2007 Green Apple Awards for Environmental Best Practice in the Green Champions category. Here they explain how they have transformed the organisation to achieve much greater environmental sustainability.

Getting started

The path to greening an organisation begins *at the top*. Gentoo's chief executive realised early the importance of environmental sustainability within the housing sector. A director has been appointed to head up Gentoo Green alongside two board members, with the sole purpose of building a culture in which environmental sustainability matters.

The road to winning hearts and minds begins with an organisation's *capacity* to act, which stems from staff's readiness to take action. All members of staff were invited to put forward suggestions and ideas on how to make sustainability real in everyday work. The organisation then asked for volunteers to become environmental champions. These members of staff then formed Gentoo's very own green team and became the drivers for cultural and organisational change. A working group consisting of a cross-section of staff was set up which in one year has grown to over 40 members. From this a number of sub-groups emerged to tackle individual issues. Gentoo joined forces with Home Group to make the impact of its actions greater and to increase creativity in the process.

Taking action

In order to *identify priorities*, a consultant was appointed to identify the most significant issues and priorities for the business. From this, Gentoo established five overarching themes: *energy, waste, properties, transport* and *recycling*.

These themes not only direct work towards maximising efficiency across every part of the business but also reflect issues that are prevalent in a particular locality. For example, recycling facilities for over 2,000 people living in multi-storey flats were established in conjunction with the local authority to benefit both Gentoo's tenants and the wider community.

The EASI campaign, which provides environmentally themed information and training to tenants, promotes changes in personal behaviour to encourage more sustainable lifestyles. Gentoo also works with local schools through its Community Kids Programme, which includes an environmental sustainability session helping children (many of who live in Gentoo homes) to think about more sustainable lifestyles.

→

This last point highlights the importance of *developing partnerships*. Schools are just one of the networks which housing providers must tap into, alongside local authorities and contractors with whom best practice can be shared and sustainable practices encouraged.

Gentoo is working with the Eden Project, Sustainable Homes, Carbon Trust and the Energy Savings Trust to name a few and have developed their champions programme in direct partnership with Home Group. Gentoo recognises the importance of working with partners not only for the purpose of learning and sharing but to increase reach and impact.

Cultural change

The root to changing attitudes and behaviours within Gentoo's workplace lay in *training and empowerment*. The most influential drivers have been the environmental champions who receive specific training to start with, including a two-day residential course with Home Group, and ongoing support and guidance.

The environmental agenda has been incorporated into induction training for new starters and a half-day session has been added to Gentoo's management development programme. A green training plan complements the champions' training and promotes training opportunities available to all staff members.

The *action plan* has been crucial in taking the agenda forward as well as helping to embed environmental thinking across the organisation. It sets out three strategic aims:

- everybody makes a difference to/positive impact on the environment;
- everything that Gentoo does positively addresses the impacts on the environment; and
- Gentoo makes a difference in the sector and wider business community.

Gentoo has succeeded in getting the basics right through initiatives such as the binless office pilot and work alongside the local authority to introduce recycling facilities in their tower blocks. Yet, if these actions had not formed part of an overall plan, they would have been isolated. Of the 629 staff ideas generated, the champions took ownership of 108 and this formed the basis of an action plan, which identifies quick wins like the 'Switch it Off' campaign, 'Paper Challenge' to reduce usage/consumption and the use of grey scale/double-sided printing.

More importantly, the plan sets timescales, responsibilities, areas for development and communication methods to staff and customers. Stakeholders can see how, why and when each factor that impacts on the greening of Gentoo's business and the community in which it sits will be broached. The organisation is also in the process of seeking ISO 14001 accreditation.

→

Monitoring and communication

The final step to note on the journey to 'greening' an organisation (but not the final step to becoming sustainable) is the need to *record and review*. A year on from the launch of the champions programme the initiative has been reviewed in order to see what works and what needs to be changed. This resulted in:

- a CO_2 charter accompanied by a manager's checklist that will support it and bring consistency across the business which is currently being developed;
- publishing Gentoo's green generation page on the staff intranet which includes news articles, facts, environmental policy documents, a 'who's who' of the green team, environmental themes and links and information on the organisation's 'footprinting' service;
- a jargon buster, i.e. an exhaustive and continuously expanding tool that explains everything 'green' in simple terms;
- tool box talks for trade staff;
- an executive team challenge (each director was challenged to make a difference in their teams); and
- giving staff the opportunity to drive hybrid vehicles.

The environmental champions themselves monitor progress on an ongoing basis, alongside an Excellence Matters Environmental Sustainability Project Team, which identifies areas of success, priority and future development. Gentoo's tenants' handbook, alongside regular customer publications, ensure that customers are fully aware of how to be more environmentally friendly and how the organisation is delivering on this agenda. Staff are kept in the loop via intranet, a weekly newsletter, bi-annual staff magazine, environmental publication *Eco Express* and a monthly briefing session.

Overall, Gentoo's energy-efficiency programme has delivered over £4 million in fuel savings for tenants and reduced carbon emissions by over 7,000 tonnes per annum. The programme was run in conjunction with EDF, Powergen, British Gas and Millfold. This has not only reduced the environmental impact of the stock but also led to some shift in the behaviour of occupants.

One of the most important parts of the journey is to *celebrate success*. Attending the award ceremony for winning the increasing environmental sustainability category of the 2007 UK Housing Awards, one environmental champion commented:

> 'When I see the benefits of the group becoming more sustainable and am part of the celebrations, it gives me the motivation and momentum to continue pushing forward the sustainability agenda and encourage my colleagues to do the same.'

References

Anderson, R. (1998) *Mid-Course Correction: Towards a sustainable enterprise: the Interface model*, White River Junction: Chelsea Green Pub

Beyond Green and Consensus (2007) *Consensus Research International Sustainability Report*. Available from: http://www.consensus-research.com/articles/Sustainability%20Report%20Final.pdf

Callcutt, J. (2007) *The Review of Housebuilding Delivery – the Callcutt Review*, Communities and Local Government. Available from: http://www.callcuttreview.co.uk/default.jsp

CLG (2006) *Code for Sustainable Homes: A step-change in sustainable home building practice*, Communities and Local Government. Available from: http://www.communities.gov.uk/publications/planningandbuilding/codesustainable

CLG (2007a) *Homes for the future: more affordable, more sustainable – Housing Green Paper*, Communities and Local Government. Available from: http://www.communities.gov.uk/publications/housing/homesforfuture

CLG (2007b) *Table 577 Housing market: ratio of median house price to median income by district, from 1997*, Communities and Local Government. Available from: http://www.communities.gov.uk/documents/housing/xls/322286

CLG (2008a) *Table 209 Housebuilding: permanent dwellings completed, by tenure1 and country*, Communities and Local Government. Available from: http://www.communities.gov.uk/documents/housing/xls/323495

CLG (2008b) *Table 503 Housing market: simple average house prices by new/other dwellings, type of buyer and region, United Kingdom, from 1986*, Communities and Local Government. Available from: http://www.communities.gov.uk/documents/housing/xls/140951

Egan, J. (2004) *The Egan Review: Skills for Sustainable Communities*, Communities and Local Government. Available from: http://www.communities.gov.uk/publications/communities/eganreview

Forum for the Future (2007) *The future leaders survey 2006/7*, Forum for the Future, UCAS. Available from: http://www.forumforthefuture.org.uk/files/Futureleaders0607.pdf

GLA (2007) *Action Today to Protect Tomorrow: The Mayor's Climate Action Plan*, Greater London Authority. Available from: http://www.london.gov.uk/mayor/environment/climate-change/docs/ccap_execsummary.pdf

Grande, C. (2007) 'Wave of eco-marketing predicted', *Financial Times*, 12 February

HC (2007) *Partnering Programme Agreement*, Housing Corporation. Available from: http://www.housingcorp.gov.uk/server/show/ConWebDoc.8446

HC, WWF, Insight Investment and Upstream Strategies (2007) *Building a sustainable future: UK Home builders' progress in addressing sustainability*. Available from: http://www.housingcorp.gov.uk/server/show/ConWebDoc.12461/changeNav/440

IEMA (2006) *Land Securities – building sustainability*, Institute of Environmental Management and Assessment. Available from: http://www.iema.net/readingroom/show/16668/c189

IEMA (2008) *Sustainable Procurement at BT*, Institute of Environmental Management and Assessment. Available from: http://www.iema.net/readingroom/show/6205

InterfaceFLOR (2007) *Sustainable Futures: An Insight into Sustainability Trends in Business*, InterfaceFLOR

IPCC (2007) *Climate Change 2007: Summary for Policymakers*, Intergovernmental Panel on Climate Change. Available from: http://www.ipcc.ch/pdf/assessment-report/ar4/syr/ar4_syr_spm.pdf

Llewellyn, J. (2007) *The Business of Climate Change*, Lehman Brothers. Available from: http://www.lehman.com/press/pdf_2007/TheBusinessOfClimateChange.pdf

Marketing Green (2007) *Green Branding Imperative*. Available from: http://marketinggreen.wordpress.com/2007/04/17/green-branding-imperative/

Monbiot, G. (2006) *Heat: how to stop the planet burning*, London: Allen Lane

Smith, A. (2007) 'Taylor Wimpey merger creates UK's biggest housebuilder', *Building*, 26 March. Available from: http://www.building.co.uk/story.asp?storycode=3083761

Sponge (2007) *Eco Chic or Eco Geek: The Desirability of Sustainable Homes – Executive Summary*, Sponge Sustainability Network. Available from: http://www.spongenet.org/library/Sponge%20Eco%20Chic%20or%20or%20Eco%20Geek%20The%20Desirability%20of%20Sustainable%20Homes%20Executive%20Summary.pdf

UNEP (2007) *Global Environment Outlook GEO-4*, United Nations Environment Programme. Available from: http://www.unep.org/geo/geo4/media/

Welch, D. (2007) 'Carbon offsets – enron environmentalism or bridge to the low carbon economy?', *Ethical Consumer*, Issue 106

World Commission on Environment and Development (1987) *Our Common Future*, Oxford: Oxford University Press

WWF (2007) *Brown: hard choices ahead on climate change*, World Wildlife Fund. Available from: http://www.wwf.org.uk/news/n_0000004559.asp

CHAPTER 16:
Lessons from other European countries

Dr Minna Sunikka

Introduction

This chapter presents three case studies that offer lessons for the UK in developing policy to improve the energy efficiency of the existing housing stock. Examples have been chosen from countries with a long-standing experience in sustainable building policies, namely the Netherlands, Germany and Finland. The programmes have been selected according to their effectiveness, cost-efficiency, legitimacy and the potential to address barriers like long payback times and the lack of practical support – often identified as issues in sustainable renovation projects (Itard *et al.*, 2007; Klinckenberg and Sunikka, 2006).

The Energy Performance of Buildings Directive (EPBD) has become one of the main instruments of the European Union to improve energy efficiency of existing buildings. The policy has been encouraged by the fact that energy labelling for household appliances has proved relatively successful (Beerepoot and Sunikka, 2005). Article 7 of the directive requires an Energy Performance Certificate (EPC) for a building every time it is sold or rented (EC, 2003). Since the directive gives only recommendations and lacks any 'regulatory' teeth, its effectiveness even as a communication instrument has been questioned (Beerepoot and Sunikka, 2005; Sunikka, 2006).

Based on the experiences from the Netherlands, Finland and Germany, the case studies focus on how the use of EPCs in combination with fiscal policy instruments could work most effectively in the UK. Since costs represent the main barrier to sustainable renovation practices, the case studies focus on fiscal policy instruments:

(i) subsidised energy audits (the Netherlands);
(ii) subsidised energy audits connected to public-private performance agreements (Finland); and
(iii) preferential loans and practical support for renovation (Germany).

The information has been obtained from key policy documents, impact assessments, expert interviews and previous policy studies by the author (Sunikka, 2006; Klinckenberg and Sunikka, 2006). Policy studies are strictly time-bound and it should be noted that in this chapter the baseline years are 2005-2007.

Energy Performance Advice (Netherlands)

The Energy Performance Advice (EPA), introduced in 1996, has points in common with EPCs and is a pioneering scheme which has been extensively monitored. Subsidies were made available for energy audits and subsequent energy-efficiency improvements in the existing housing stock. The owner or landlord could ask a trained EPA assessor to make an on-site assessment of the energy performance of the building, including advice on potential improvements, detailing investment costs and potential energy savings. Compared to the implementation of EPCs in most countries, the EPA was labour-intensive, costing around €200 for a single-family home.

The EPA was part of a wider policy – the Energy Premium Regulation (EPR), which included a Regulatory Energy Tax (RET) that was connected to a certificate-based subsidy system to improve energy efficiency of private households. During 2000-04, energy premiums were made available to subsidise the purchase of A-rated energy-efficient household appliances and energy saving measures recommended in the energy certificate. Appliances could be purchased at *any* shop and consumers would later receive the subsidy on handing in the receipt. The energy premiums were paid by energy distributing companies, who could deduct them from the RET they had to pay on the energy sold to small-scale users. It was expected that subsidies for energy-efficient products and energy audits, coupled with increasing energy prices would encourage households to invest in energy improvements and, since the RET did not apply to renewable energy sources, that the green electricity market would grow further.

The programme did help to disseminate information on home energy improvements and energy-efficient appliances among consumers. The number of assessed houses increased to 300,000 in 2003, of which the vast majority were rented properties, and owned by large housing associations (75 per cent of all cases). As an additional benefit, housing associations often integrated EPA assessments in their environmental policies (Sunikka and Boon, 2002). The number of assessed owner-occupied homes (initially the target group of the policy), was below target (27,000 in 2003). Altogether, the energy measures taken following an assessment, equated to a 0.4 Mt/year reduction in CO_2 emissions. The average cost to the Dutch government was €200 per tonne of CO_2 saved. The administrative cost of the programme was high, given the relatively small energy savings per application (Harmelink *et al.*, 2005).

Almost 75 per cent of home owners indicated that the EPA advice had not changed their planned investments. 89 per cent of owner-occupiers, who asked for advice and decided to invest in energy-efficiency measures as a result of it, had already planned to do so. The number of such 'free-riders' was a result of the structure of the programme. In the first years, owners only received a subsidy if they had implemented at least one recommended measure, however, later subsidies were also given for the advice itself. If owners implemented more than one of the recommended energy improvements, they would receive a 25 per cent higher subsidy (later ten per cent). The selection of measures that qualify for funding is a key issue if an energy certificate is to be coupled with subsidies and many of the measures eligible for EPA funding had already been standard practice in the industry before it was introduced (Klinckenberg and Sunikka, 2006).

Whilst around 25 per cent of the housing associations that asked for advice had consequently changed their investment decisions, 75 per cent indicated that the advice had little or no impact on their decision to invest in energy-efficiency measures. Tenants at the time were not asking for energy-efficiency measures.

Subsidies for EPA were abolished in 2004. Officially, this was due to the free-rider effect but at the time there was a political shift away from environmental priorities. The scrapping of the programme gave a bad signal to consumers. The tax on household energy consumption that supported the EPA had only limited effects. Despite its huge impact on energy prices at the time it was introduced, research shows that only half of the Dutch population was aware of the RET, and only two per cent took it into account in their use of electricity (Van der Waals, 2001). Whilst fuel poverty numbers in the Netherlands are lower than in the UK, households living in badly insulated homes already pay higher energy bills, and thus suffer more from taxes, without having access to low-interest loans to invest in energy efficiency. Despite the free-rider effect, which could be reduced by changing the structure of the programme, subsidies were a more successful part of this composite policy than the RET tax. By itself it would have had a very limited effect in supporting the energy improvements recommended in the certificates.

Energy saving agreements and subsidised energy audits (Finland)

Despite its young building stock (mostly built following the energy crisis in the seventies), Finland recognises the potential of existing housing to deliver energy savings and recently introduced a policy for sustainable renovation (YM, 2007). In the Finnish National Climate Strategy and the associated Energy Conservation Programme, voluntary energy conservation agreements play a central role. Energy conservation agreements are public-private framework agreements made between the Ministry of Trade and Industry and organisations from various sectors, one of them the real estate sector. The agreement is voluntary but financially supported and monitored by the government. Every €1 of public investment saw a €5 return in reduced energy consumption (Heikkila *et al.*, 2005).

Participating companies receive 40 per cent support for their energy audit costs, 15-20 per cent support for energy improvement investments recommended in the audit (up to 40 per cent for new technologies), and practical support in setting up an environmental management system (ISO 14001) which it is hoped will initiate more environmental investments. In the period 1996-2003, the government invested a total of €16.1 million in energy performance agreements, both for energy audits and energy-efficiency investments.

Overall, the experience from the performance agreements and energy audits has been positive. The number of buildings undergoing energy audits has increased remarkably. Most participants stated in the monitoring report that their energy consumption had already reduced due to the installation of measuring equipment and monitoring practices.

The Finnish programme is characterised by wide coverage involving various sectors of the economy, with active participation in each sector. There is a focus on specific energy saving actions and practical assistance is available for the implementation of the agreements led by non-profit energy agency, Motiva. At the time of writing, the Ministry of Trade and Industry is consulting stakeholders on a new round of contracts, covering the period 2008-2016. The aim is a nine per cent reduction in energy consumption by 2016, which the real estate sector has signed up to.

This programme, however, also carries the risk of a free-rider effect. Not many participants have acted on the audit and there was lack of awareness among respondents as to whether they would actually qualify for an investment subsidy (many were under the impression that it would extend to ground-breaking technologies only). Many respondents remarked that audit subsidies were the main reasons to join the agreement. One barrier to more large-scale improvements, particularly in the rented sector is the so-called 'split incentive' effect – a landlord signing up the agreement has no incentive to invest in energy improvements since tenants pay for the energy used. It also appeared that if the property was likely to be sold in the near future, there was no incentive to have it audited. This was also the case with rented properties in particular and landlords with changing real estate portfolios, where investments in energy-efficiency improvements were unlikely. Shared ownership properties were not audited at all.

In the real estate sector, the agreement covered 12.8 million m² of property, with 82 per cent of the properties participating in the agreement owned by members of the Association of Real Estate Developers in Finland (RAKLI). The target was to reduce heating energy consumption of these buildings by ten per cent in 2005 compared to 1990, and by 15 per cent in 2010. The targets, however, have proved to be too ambitious. In 2004, heating energy consumption (kWh/ m²) was reduced by 3.4 per cent compared to 2000. Average electricity consumption on the other hand stabilised. Water consumption was reduced by 20 per cent during the same period. Yet on the whole, the real estate sector has delivered less savings than other sectors.

A review also revelaled the labour intensity of the programme in the real estate sector. The sector had by far the highest investment costs, averaging at around €11,000 for each party. This included work input and other costs. In relation to energy savings, this amounted to €37.6 per MWh/a. Financial support by the government focused on energy audits whereas implementation of the suggested energy improvement measures had been slow. The proportion of financial support in relation to the actual energy saving gains amounted to €28.3 per MWh/a in 2003.

Participants felt that some of the targets in the agreements were too vague and difficult to monitor, such as '...*surveying the R&D in energy efficiency*' (Heikkilä *et al.*, 2005). The simplest targets on the other hand worked best (e.g. measurement of energy consumption). Various objectives that might have been included, such as separating energy costs in the rent payable or educating residents about energy use were left out. This again supports an approach based on specific reduction targets, such as adopted in the German case study (see below).

A major shortcoming of the current programme is that the estate portfolio of the parties in the agreement (mainly professional landlords) is constantly changing. In the current agreement, the signed parties did not have to include all their stock but could choose the properties they wanted to be monitored during the process. It was observed that some parties played safe and presented properties for the monitoring that were energy efficient anyway. They were not necessarily the same as when the agreement was signed. Yet the point was to encourage energy-efficient improvements, targeting the less efficient stock. It has been suggested that in the future, the specific projects that will be reported for the agreement should be fixed when the agreement is signed. If the properties are not specified, it is not possible to give an accurate picture of the energy improvements as the owner may move towards a more energy-efficient portfolio anyway.

Experience from this programme shows the variety of Environmental Impact Assessment (EIA) systems available in the market. Although the number is increasing, there is still no sign of a standardised EIA approach. In Finland, while some of the properties participating in the agreement had already been ranked using one of the EIA systems, no links were made between them and the energy audit results. This oversupply of methods and tools is a dominant feature in countries like Finland (e.g. PromisE, EcoProP, PIMWAG), the Netherlands (e.g. EcoQuantum, EPA, GreenCalc) and even the UK (e.g. BREEAM, ENVEST), where a lot of investment has gone into R and D and the development of environmental management tools.

KfW CO_2 reduction programme and Energiepass (Germany)

Germany is one of the first EU countries to adopt a more mandatory approach in tackling the energy efficiency of its existing stock. The new German building regulations stipulate that if more than 20 per cent of the building component area is changed, it has to be done in line with the regulatory requirements for new construction (Sunikka, 2006). For instance, if more than 20 per cent of the window area needs replacing, building regulations for new build will apply. The implementation of mandatory requirements in the existing stock has not been without problems though, and actual energy saving figures as a result of the new regulations are not yet available.

There are, however, two voluntary programmes for the existing stock that may be of interest to UK policy-makers, namely the CO_2 reduction programme of the German Kreditanstalt für Wiederaufbau (KfW) and the Energiepass.

The KfW CO_2 reduction programme

In this programme the Kreditanstalt für Wiederaufbau (KfW) offers loans at low-interest rates for renovation projects that meet the CO_2 reduction target of 40 kg CO_2 per m^2 (Kleeman et al., 1999). The fact that a specific target is required instead of using a general list of energy measures or products, allows for flexibility in implementation, raises knowledge and awareness of home energy efficiency (building owners are required to prepare a fairly precise assessment of the energy performance of a building for a loan application) and enables quantitative monitoring of the programme. If a similar programme of loans with low-interest rates for energy-efficient renovations were to be implemented in the UK, it could also be structured in a way that energy cost savings could be directly used to repay the loan.

Investment of around €3.2 billion has enabled 166,600 dwellings to be renovated since 2000 (Klinckenberg and Sunikka, 2006). These measures achieved a reduction in CO_2 of between 2-2.5 Mt by 2005, less than half the originally anticipated CO_2 savings set out in the National Climate Protection Programme. The total budget was recently increased to €1.4 billion per year, of which €800 million for preferential loans. Since 2001, the programme has created up to 25,000 jobs related to energy-efficient renovation, and a further 12,300 related to new-build improvements each year. Together with the climate protection programme for existing buildings, total CO_2 savings of between five and seven Mt per year were expected in 2005. Some of the expectations, however, have been over-optimistic, with policy-makers tending to over-estimate outcomes when under pressure to deliver overall targets (Wagner et al., 2005). The programme has also been weakened during implementation. In the design of a similar policy programme it should also be considered that handling the applications is labour-intensive and in order to achieve adequate savings the CO_2 reduction target needs to be set high enough (yet it should be in proportion to the costs of the improvements required to achieve it). As with the Dutch scheme, there is a risk of a free-rider effect where a loan is used by people who would have implemented energy-efficiency measures anyway, and this aspect has not been taken into account in the monitoring process.

Chance Energiepass

The second example that could address some of the barriers to energy-efficient refurbishment in the UK is the Chance Energiepass programme, a public-private partnership involving the German Energy Agency and private parties, such as Rockwool and BuildDesk. The programme provides home owners with a building certificate that is compatible with the EPCs in Germany. It consists of an internet tool which can be used by professional home owners (including housing associations and professional managers) for their own use, and can as well be used by DIY stores to advise their customers. The tool provides an energy rating for the building and advice on how to improve its energy performance. The Energiepass programme is characterised by several levels of advice. Each level is marked by further involvement of experts, with a corresponding price tag. Local partners can provide detailed advice about the improvement options, supported by national partners who provide marketing materials, a consumer website, and technical support.

In this way, the programme addresses the barrier of practical support that a lot of non-professional owners encounter. Via the website, consumers can assess additional information, such as the available financial support options in their region. Local partners, mainly DIY stores, are trained in building performance and the use of building energy performance calculation software. It is too early to assess the success of the programme, but in 2005/2006 more than 150 Chance Energiepass partners joined the programme and employees of these partners have been trained to become energy advisors in the building materials industry (Klinckenberg and Sunikka, 2006). This kind of programme could improve the impact of the energy certificates in the UK, while at the same time reducing the cost of obtaining one.

What can we learn from these examples?

Compared to the case study countries, the UK has an outdated building stock which presents a major challenge to achieving greater energy efficiency. At the same time, it has one of the most ambitious policies for new homes anywhere in Europe, requiring zero-carbon standards by 2016. The outdated stock means greater potential in energy savings but also higher refurbishment costs. A composite policy of energy certificates and fiscal instruments that can support the adoption of energy-efficiency measures as recommended in the certificates is, therefore, key in reducing the carbon footprint of the UK stock. With the roll out of the EPCs to be completed by October 2008 (i.e. in the social and private rented sectors), it is very timely to look for lessons from countries with a longer background in sustainable renovation policies. The case studies all demonstrate how EPCs could be enhanced by creating appropriate fiscal instruments.

The main lessons to be learned from the Dutch EPA programme are that subsidies can increase and speed up the adoption of energy assessments. However, in order to reduce the number of free-riders claiming subsidies for improvement work which had already been planned and thus threatening the economic viability of the programme, it should focus on more ambitious technologies and products. Since professional landlords tend to be over-represented among applicants in programmes like the EPA, any potential initiative in the UK should pay special attention to reaching private households. Equally, the continuation of any programme should also be assured. In the Netherlands for instance, the scheme was brought to a halt once environmental issues slipped off the political agenda. The technical structure of the EPA programme could provide valuable lessons for UK policy-makers. While the programme is more in-depth and labour-intensive than the implementation of the energy certificates implemented as a part of the Energy Performance of Buildings Directive (EPBD) in the UK, consumers are provided with more meaningful information than a standard label whose feedback is very limited and standardised.

The Finnish approach of using voluntary energy performance agreements within different sectors, and leaving them to choose the most cost-effective way to achieve the agreed targets, could fit the political climate in the UK which generally favours market-led solutions. The Finnish experience further highlights the need to include concrete targets in the agreement. If a similar programme were to be implemented in the UK, it would be important to include single-family houses, which account for almost 50 per cent of space heating energy consumption in Finland but are outside the scope of publicly supported energy audit programmes. Finding ways to persuade professional landlords to invest in energy efficiency would further improve the effectiveness of the programme. For instance, landlords likely to sell off stock are unlikely to invest. Equally, if dwellings are not specified in the agreement, owners may select those for the monitoring process that fulfil the criteria anyway.

The German KfW CO_2 reduction programme, offering loans with low-interest rates for energy-efficient improvement schemes, could work well in the UK. Preferential loans might be combined with the EPC classifications as well as mandatory requirements. Under this scenario, preferential loans would be approved on the basis of moving the property into a higher energy rating band (e.g. from D to C). The strength of the German programme is a clear CO_2 reduction target per m², which allows flexibility as to how this is achieved. The programme is also responsible for creating new jobs in the construction industry, benefiting small and medium-size firms in particular.

The main challenges in all three case studies are the free-rider dilemma, the capture of benefits in the rental sector and the gap between the energy audit recommendations and their implementation. The free-rider dilemma tends to affect all programmes with either direct or indirect subsidy regimes. A specific reduction target (KfW CO_2 reduction programme) may result in reduced numbers of free-riders compared to a programme that uses a list of subsidised measures (Dutch EPA). The proportion of free-riders is usually higher among professional landlords with a large portfolio than private households, particularly those on low-income who are more likely to occupy energy-inefficient housing. In both the Dutch and Finnish programmes, professional landlords made up the majority of those claiming energy audit subsidies.

Compelling private landlords to improve the energy efficiency of their properties remains a major challenge as long as tenants pay the energy bills and market demand for high energy performance is lacking. This is likely to continue to be the case, especially in areas where housing demand constantly exceeds supply. The way to compel professional landlords to make large-scale energy-efficiency improvements seems to point towards a more mandatory approach. However, this is not without problems – as can be seen in Germany where regulatory requirements for large renovation schemes have been introduced.

Despite success in terms of the number of properties audited in all three programmes, there was less enthusiasm shown in implementing the recommended improvements. One reason for this is the prospect of a property sale in the rented sector which affects the owner's willingness to make the required investment. Therefore, with any programme, the real challenge lies in implementing the recommendations given to owners. In the owner-occupied sector, a lack of financial support for the measures to be taken, as well as the difficulty of choosing appropriate products and contractors, remain as major barriers that cannot be overcome with financial support alone. The German Energiepass programme, where a client can get further advice and support from a local DIY store once an energy certificate has been issued, is an interesting initiative which merits further exploration in the UK context.

The case studies also highlight the gap between the effectiveness of current policy instruments and targets. In the successful KfW CO_2 programme for example, the original targets were too ambitious. In this context, the UK's target of a minimum of 60 per cent reduction in CO_2 emissions seems to be very ambitious. As seen in the case studies, it is challenging to even stabilise the increase in current energy consumption, let alone to create demand for measures to reduce consumption. If considerable CO_2 reductions in the housing sector are to be achieved, a move towards a more mandatory approach needs serious consideration.

But in the current housing market where demand exceeds supply (especially in the UK), it is difficult to justify more stringent regulations without government support.

The combination of building regulations with the energy labels associated with the EPCs is an interesting approach that merits further consideration. Energy regulations, however, cannot be imposed on the existing housing stock overnight as most energy measures are not yet cost-effective and not all households are in a position to comply with mandatory standards. This is further exacerbated by the high levels of fuel poverty in the UK.

Compliance also remains a key issue. In the case of Denmark, for instance, (which has a mandatory energy certificate scheme that is usually considered as a prototype of the EU energy certificate), only 50-60 per cent of the potential buildings are registered in the scheme despite it being mandatory in principle, as there are no monitoring systems or sanctions for non-compliance. There is potential to increase compliance with information campaigns as 43 per cent of respondents had never heard of the scheme. Acceptance of the scheme was relatively high, but many building owners were not aware of the certification requirements, which tend to get buried under all the paperwork involved when a building is sold (Laustsen, 2001).

Yet, legislation could produce better policy outcomes if compliance and legitimacy are ensured, the 'dilemma' of low-income households is addressed and the behaviour of occupants does not create rebound effects.

There are also different kinds of regulatory policy instruments that could be more useful in the UK context than traditional building regulations – for example, having above standard requirements for government buildings. Already used in some states in the US, the UK government has set a target for all public buildings to be zero-carbon by 2018 (Planning, 2008). Other regulatory instruments could include a speeding up of the building permit process as well as giving extra surface area as permitted construction density for environmentally high-performing projects.

Because of national differences in housing stock and policies, it is would be better if the EU had a more strategic role – influencing the policies of national governments at a strategic rather than operational level. EPCs have the potential to be the 'baseline tool' for policy-makers across the EU to improve the energy efficiency of the existing housing stock. For this reason the case studies presented here look at how to improve the energy certificates rather than suggesting something completely new altogether.

A more mandatory approach adopted by the government, supported by fiscal instruments, would bring the UK to the forefront of energy-efficiency policy in the EU. This role may be appropriate for the UK, given the increasing importance of climate change within the media, society and the political arena.

References and further reading

Beerepoot, M. and Sunikka, M. (2005) 'The contribution of the EC energy certificate in improving sustainability of the housing stock', *Environment and Planning B*, 32 (1) 21-31

EC (European Commission) (2003) 'Council Directive 2002/91/EC of 16 December 2002 on the energy performance of buildings', *Official Journal of the European Communities,* No L 1, 04/01/2003, pp. 65-71

Harmelink, M., Joosen, S., Blok, K. (2005) 'The theory-based policy evaluation method applied to the ex-post evaluation of climate change policies in the built environment in the Netherlands', in S. Attali and K. Tillerson (Eds.), *ECEEE 2005 Summer Study Proceedings*, Stockholm: ECEEE

Heikkilä, I., Pekkonen, J., Reinikainen, E., Halme., K. and Lemola, T. (2005) *Energiasopimusten kokonaisarviointi*, Helsinki: Ministry of Trade and Industry

Itard, L., Meijers, F., Vrins, E. and Hoitings, H. (2007) *Building renovation and modernization in Europe: State of art review*, Delft: OTB

Kleemann, M., Kuckshinrichs, W. and Heckler, R. (1999) 'CO_2-Reduktion und Beschäftigungseffekte im Wohnungssektor durch das CO_2-Minderungsprogramm der KfW: Eine Modellgestützte Wirkungsanalyse', *Schriften des Forschungszentrums Jülich, Reihe Umwelt Band*, 17, Jülich

Klinckenberg, F. and Sunikka, M. (2006) *Better buildings through energy efficiency: a roadmap for Europe*, Brussels: Eurima. Also available at: http://www.eurima.org/downloads/EU_Roadmap_building_report_020307.pdf

Laustsen, J. H. (2001) 'Mandatory labelling of buildings: the Danish experience', *Sustainable Building* 4, pp. 12-14

Planning (2008) 'Date set for zero carbon', *Planning*, Issue 1760, 14 March

Sunikka, M. and Boon, C. (2002) 'Environmental policies and efforts in social housing: the Netherlands', *Building Research and Information*, 31 (1), pp. 1-12

Sunikka, M. (2006) *Policies for improving energy efficiency in the European housing stock*, Amsterdam: IOS Press

Van der Waals, J. F. M. (2001) *CO_2 reduction in housing, Experiences in building and urban renewal projects in the Netherlands*, Amsterdam: Rozenberg

Wagner, O., Lechtenböhmer, S. and Thomas, S. (2005) 'Energy efficiency – Political targets and reality, Case study on EE in the residential sector in the German Climate

Change Programme', in S. Attali and K. Tillerson (Eds.), *ECEEE 2005 Summer Study Proceedings*, Stockholm: ECEEE

YM (2007) *National Renovation Strategy 2007-2017*, Linjauksia olemassa olevan rakennuskannan yllapitoon ja korjaamiseen, Helsinki: Ymparistoministerio

CHAPTER 17:
The debate on energy and climate change – a different perspective

John Perry

Introduction

Daily power cuts, dramatic increases in fuel costs, public buildings closed half the day to save energy, generating plants unable to operate, water shortages caused by lack of power to work the pumps, hikes in food prices driven in part by escalating transport costs. Does this sound like a forecast of an energy-starved future in a few decades time? In fact, it's the present-day reality in the small Central American country of Nicaragua.

Like many other developing countries, Nicaragua has tremendous potential to develop renewable energy sources. Yet the irony is that poverty forces it to be even more of an oil economy than most developed countries. Electricity generation depends 85 per cent on burning bunker fuel and, with no railway system, motorised transport is 100 per cent oil dependent. This makes it crucially vulnerable to changes in energy markets.

This chapter aims to give a different perspective on the issues discussed in the book, by looking at a country which in many respects is at the opposite extreme from Britain in terms of climate, resources and geography. So why include this different perspective? What's it got to do with how things might change in Britain?

First, there are some points of comparison, and perhaps some lessons which can be drawn about Britain's future, looking at Nicaragua's present. But more generally, Nicaragua is an example of what is happening in terms of energy and climate change in poorer, developing countries, and in many respects this directly links to what is happening in richer countries like Britain. Obviously, we all share the global climate, and the effects of changes driven by energy consumption in northern countries are being felt just as acutely – or more so – in southern ones. When we aim for 'fuel security' for Britain, we might achieve it at the cost of greater insecurity for people elsewhere. And when we try to reduce fuel poverty, do we consider the extremes of fuel poverty experienced by people who have no access to formal energy systems at all?

This could be an ambitious agenda for one chapter of a book, and rather than attempt to do it full justice I will provide some examples from this one small country, and signpost sources for readers who want to look at the issues in more depth.

Some basic facts and figures

Nicaragua is 11-13° north of the equator, enjoying a warm climate with little seasonal temperature change. Its land area is the size of England and Wales, yet its population is less than six million. Most of those people live on the drier, Pacific slope of the country, in a region dominated by a chain of active volcanoes. In the sparsely-populated eastern half, broad rivers drain to the Caribbean, often through still-virgin rainforest. The marked six-month dry season of the Pacific coast is much less obvious further east, where most months have some rainfall.

Nicaragua exploits no fossil fuel of its own, and in fact is a modest carbon 'sink' for the rest of the planet, having the largest area of rainforest in the western hemisphere, north of the Amazon. As we shall see, its potential for generating power from renewable sources – solar, geothermal, wind, hydro and even biofuels – is enormous but only slowly being tapped. The biggest source of 'renewable' energy is burning wood, which accounts for half of total fuel consumption. But of course, whether it is renewed or not depends on conservation of the forests. Even counting firewood, though, Nicaragua is dependent on imports for almost half the energy it consumes.

This book focuses on housing and on domestic energy consumption, and mitigating the impact of climate change. This chapter will therefore concentrate on Nicaragua's domestic sector, focusing on consumption of electricity and firewood, and largely disregarding transport. It will look at how and why Nicaragua has reached the situation it is now in, and at alternative models for change. It will look at two examples at local level – one which has worked and one which hasn't. And it will offer some comments, from the perspective of a small developing country, on the issues of energy and climate change in the rich countries of the north.

A bit of history

Like people of other developing countries, Nicaraguans of a generation or two back were largely self-sufficient in energy because they used so little and what they used came from local sources. In 1950, over 60 per cent of Nicaragua was covered by tropical forest. With a small and largely rural population, using firewood was probably completely sustainable. Now though, the forest has reduced to just over 40 per cent: still a significant figure, but of course the 'agricultural frontier' has pushed further and further away from the cities on the Pacific coast, making the major forests more remote. Firewood for the majority of the population is now, at best, only a partially renewable resource.

Energy dependency started to grow as the electricity grid developed in the 1950s. Nicaragua has no coal so the obvious choice of fuel for power stations seemed to be oil – which at the time was being produced in abundance in various parts of the Americas. Ironically, the right-wing dictatorship which was to collapse in 1979 built a largely publicly-owned electricity system and eventually invested in the first hydro-electric plants. It also initiated a geothermal plant to start to make use of heat from the volcanoes.

In the revolutionary 1980s, the country's priority was to survive the war against the 'Contra' (counter-revolutionaries) and a damaging blockade by the United States: investment in energy came to a halt. The Sandinista government, legitimised in elections in 1984, was replaced by a series of right-wing governments, starting in 1990 and running until the beginning of 2007. In the 1990s, if there was a 'strategic vision' for the energy sector it was to privatise both the generators and the distribution system, make even greater use of oil, and trust the private sector to start to invest in a more modern system. The geothermal plant was privatised and the hydro-electric plant allowed to run down to the point where it operated well below capacity.

Although most urban areas had electricity, in 1990 half the population were still not connected to the grid. These were predominantly people in rural communities but also informal settlements adjoining towns. With the overall neglect of rural needs in the 1990s, there was little change. (Since 2000, little progress has been made, often through off-grid initiatives, discussed later.)

The pressure to privatise

As is now well known, the economic model developed in the USA and Britain in the 1980s was later imposed on developing countries by the international financial institutions – the World Bank, the International Monetary Fund (IMF) and (in Latin America) the Inter-American Development Bank (IDB). The model advocated cuts in public spending and the privatisation of state-owned enterprises, especially utilities such as energy companies.

Nicaragua is a small and poor country with, until recent debt-relief initiatives, one of the world's highest levels of international debt (around $1,000 per head of population, in a country where GDP is still only $1,038 per head). Nicaragua could ill-afford to challenge the impositions of these international finance institutions. But in any case, for 16 years it had governments that embraced them enthusiastically and were able to portray them to critics as inescapable requirements.

By 2000, almost all the generators had been privatised. Only public opposition prevented the hydro-generator, too, being sold off at a knock-down price in 2002 (for only $40 million, when it was still producing annual profits of $18 million). As happened in Britain, as an enterprise left isolated in the public sector it then paid the penalty – it was starved of cash and even had its revenues raided to subsidise the private operators.

The crown jewel of privatisation was the electricity distribution system, which was supposed to be sold off competitively (as happened in Britain). However, only one company – the Spanish-based multinational Union Fenosa – made a bid. It created two distribution companies and various offshoots to give the appearance of not being a monopoly, and bought the whole system with its bid of only $115 million (slightly more than half the value placed on it by the government). As well as the infrastructure, it acquired Nicaragua's limited technical expertise – the engineers and economists who run the system, and even the laboratory which tests household meters.

Nicaragua therefore entered the current century with a newly privatised electricity system covering only half the population, massively dependent on oil imports, and charging the highest consumer prices of any country in the region. The government had limited regulatory ability to control a multinational company whose contract was, in any case, underwritten by World Bank guarantees. It had also completely failed to develop alternative, more sustainable power sources, which might have increased its flexibility in dealing with the Spanish multinational.

Ironically, the expected benefits of the takeover by Union Fenosa are even now spelt out on one of the World Bank's websites as including reduced losses in the distribution system, extended rural electrification, and the economic advantages of more reliable electricity services and bigger tax revenues.

Nicaragua's twenty-first century energy crisis

Almost as soon as Fenosa took over in 2000, it started to come under criticism for frequent power cuts. Despite the promises, privatisation brought little new investment, so not only was power generation based on imported oil, it took place in old and inefficient power stations afflicted by frequent breakdowns. A combination of mismanagement and large numbers of illegal connections meant that, of the power that was generated and passed to Fenosa, some 30 per cent was still being lost before reaching legitimate end users. Connections to other Central American countries, intended (like Britain's with France) to import power to meet emergency demands, were often out of action.

Of course, even in the 1990s it had been obvious to intelligent observers that demand for oil was outpacing supply, and that the peaks in oil prices that had inflicted worldwide damage in the 1970s might well return as a long-term trend.

While higher oil prices in Britain affect transport costs and might lead people to curb optional travel or switch to public transport, in Nicaragua the effects of oil costing over $120 per barrel have been much more dramatic. The cost of basic foods has increased by up to 40 per cent, and the daily bus journey to work might consume most of your weekly earnings.

The effect on electricity generation as been even worse. Shortages of oil and breakdowns in the generating plants have combined to produce ever more drastic power cuts. The generating companies are charging Union Fenosa more for the power it receives, yet Fenosa finds it increasingly difficult to pass the costs onto consumers, who often already find it impossible to pay higher bills and might prefer to be cut off and perhaps reconnect illegally. As a consequence, Fenosa now owes as much as $30 million to the generating companies.

Despite the fact that the original sale had factored in guaranteed profits levels, and had even assumed a high level (15 per cent) of 'technical losses' in the distribution system, Fenosa has persistently sought higher prices or has added extra charges to bills. Also, as it installs new meters, households find that their bills mysteriously rocket. Officially, electricity prices in Nicaragua are higher than anywhere else in Central America. Unofficially, the gap is even bigger.

In 2006 and 2007, amidst widespread protests and a situation described by one of the main national newspapers as living '...*in a state of catastrophe*', the power cuts were extended and formalised so in most places they lasted up to ten hours per day. The effect on the economy was disastrous. Government offices were reduced to working half-days only. It became a common sight to see, hear and smell diesel generators sitting on pavements outside shops, keeping the lights going inside. Businesses dependent on refrigeration either stopped operating or had to invest in expensive alternative systems.

Yet these problems only directly affect half the population, as the other half continue to have no connections to the grid. The system has hardly grown at all since Fenosa took it over. Although its bid contained promises of investment, there was no contractual timetable about how and where this should take place. In fact, as we shall see, where rural areas have secured electricity for the first time, this has been through international aid rather than as a result of investment by Fenosa.

As a report by Christian Aid on the effects of electricity privatisation in Nicaragua concludes:

> '*It would require a combination of a watertight, well-defined contract at the outset, a strong regulatory regime and financial transparency to safeguard the consumers' interest in this situation...none of these are present*' (McGuigan, 2007).

Climate change worsens the problems of the poorest people

As if these problems were not enough, poorer people in Nicaragua are already suffering the effects of climate change. Nicaragua has suffered two major hurricanes in the last decade, as well as less severe weather-related disasters such as droughts and floods. Almost certainly, these unusual weather events are related to climate change provoked in large part by the way we in the north have consumed fossil fuels.

At the very least, climate change has added to food price inflation and crop failure, at the worst it has led to the devastation of Hurricane Mitch in 1998 (which killed 11,000 people across Central America) and, at the end of 2007, Hurricane Felix which hit the sparsely populated but cripplingly poor Caribbean coast, felling hundreds of thousands of hectares of forest and destroying communities. It is perhaps not surprising that Nicaragua has one of the highest outward migration rates in Latin America: at any one time more than one million Nicaraguans are working temporarily or permanently, legally or illegally, in adjoining countries or in the United States.

Could it have been otherwise?

The main theme of this book is environmental sustainability, particularly against the background of climate change and diminishing supplies of scarce resources. To this must be added dimensions of social and political sustainability – whether individual households have secure energy supplies, that as far as possible they control, and whether the same applies at national level.

In Britain, public attention is turning to these issues partly because of growing concern about climate change, but also because of the growing vulnerability of a large, developed economy to shifts in the world supply of fuels such as oil and gas. Even so, the prospect of turning on the light switch to find that nothing is working is not something that troubles most people: energy security is taken pretty much for granted.

As we have seen, in a poor country like Nicaragua, even the rich might find that the lights have gone out, and the poor may have no light at all. With the benefit of hindsight, it seems incredible that making electricity supplies secure, and available to everyone, have not (until recently at least) been the dominant goals of energy policy.

Yes this need not have been the case. If the same preparation had gone into privatisation in Nicaragua as happened when electricity was privatised in Britain, sufficient safeguards might have been built in, competition might have been secured, or the regulatory system made robust enough to adapt to changing needs. Advantage could have been taken of private capital to extend and improve the distribution system.

Furthermore, while in the 1990s there was little or no interest in looking for energy sources other than oil, it was already apparent that Nicaragua had enormous potential supplies of renewable energy. But perhaps a decade ago it was expecting too much for this to have been taken seriously into account? Had it been, requirements could have been introduced to make greater use of renewables, or to promote energy-saving measures, as happened in Britain.

But of course, whatever the rhetoric, it has been apparent that in Nicaragua (as in Britain) the principle aim of the electricity multinationals is expansion and profit generation. Union Fenosa's profit target for its worldwide operations in 2007 was €1.1 billion, and its operations now extend to eleven countries, as well as Spain.

The fundamental problems – energy security and energy poverty

Two fundamental challenges are faced by all poorer countries – and of course are echoed in developed ones – energy security and energy poverty. Both relate to and have a bearing on climate change.

Energy security

The idea of what constitutes 'energy security' is changing as its importance grows. The basic definition by the International Energy Agency (IEA) is that energy supply is 'secure' if it is adequate, affordable and reliable. The European Union has a similar definition. But it can also be widened to refer to avoiding dependence on one source of energy, or on sources whose consumption worsens climate change. As the IEA says:[1]

> 'The populations of many, especially small, oil-importing developing economies are faced with insecure, inadequate, barely affordable and unreliable energy supplies that undermine economic development.'

In the case of poor countries like Nicaragua, it is perhaps best seen at two levels: security for households and businesses, and security at national level. Many households and businesses are highly energy-insecure because in Nicaragua they rely on an electrical system which is not performing properly. But higher income households can maintain their energy security by buying generators and affording higher petrol prices.

At national level, we have seen that the overwhelming dependence on imported oil as the main energy source has created intense energy insecurity. Not only have there

---

1 ʻOlz, S. *et al.* (2007) *Contribution of Renewables to Energy Security.* IEA.

frequently been shortages 'at the pump' or (in the case of bunker fuel) 'at the generating station', the escalating prices have been extremely damaging to the economy. In 2006, Nicaragua spent $717 million on oil, which is equal to 66 per cent of its earnings from exports. As those lobbying on international debt issues have pointed out,[2] this is typical of countries like Nicaragua which are also supposed to be benefiting from the 'HIPC' (highly indebted poor countries) initiative to reduce their debts and create resources to tackle poverty. Furthermore, as international oil prices continue to rise, and demand for oil tests the limits of supply, Nicaragua's recent crisis might turn out to be the prelude to much harsher and longer shortages.

Energy poverty

In Britain, as Chapter 6 points out, fuel poverty is defined as the proportion of people who spend more than ten per cent of income on energy. However, there is a wider concept of 'energy poverty', which is about access to energy of the right type: electricity for lighting and basic domestic appliances (e.g. a radio), and more efficient forms of fuel for cooking – other than biomass (wood, animal dung, etc).

Across the world, 1.6 billion people – one quarter of the population – lack access to electricity, and no less than 2.4 billion rely on biomass. The World Energy Review 2002 showed that both categories correlate strongly with the 2.8 billion of the world population that live on less than $2 per day. Nicaragua, as the second poorest country in the Americas (only Haiti is poorer), has half its population in these categories.

As well as type of energy, the level of consumption also correlates with income. So, for example, Nicaraguans consume, on average, only one-tenth of the energy used by people in the UK. But this crude figure hides a wide spectrum. Figure 17.1 illustrates typical household fuel 'transitions' that occur worldwide as income increases and greater access is available to modern energy services and equipment to use them. The 50 per cent of 'energy poor' Nicaraguans, concentrated in rural areas, are represented by the box at the bottom left of Figure 17.1. Their energy consumption mainly means burning wood, using candles or small amounts of kerosene, and buying batteries for torches. For town dwellers, most of whom would be in the middle column of Figure 17.1, it might mean using electricity and (bottled) propane gas, taking buses or taxis, or even having a car. For a tiny elite, represented by the right of Figure 17.1, it means cars, air-conditioning, washing machines and possibly the full range of energy-intensive devices and ICT systems found in many British homes – and a level of personal consumption comparable to those in developed countries.

---

2   Jubilee USA Network (2006) High Oil Prices: Undermining debt cancellation and fuelling a new crisis? Policy Brief (available from: www.jubileeusa.org).

Figure 17.1: Illustrative example of household fuel transition

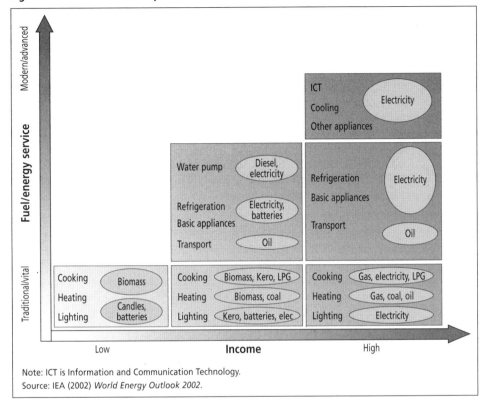

Note: ICT is Information and Communication Technology.
Source: IEA (2002) *World Energy Outlook 2002*.

Tackling energy poverty – two case studies

But let's break off from looking at the broad picture to look at two recent examples from the same Nicaraguan city of Masaya, which represent different ways of trying to end energy poverty and create energy security at community level. Both were projects to bring electricity for the first time to rural communities unconnected to the grid. The first, called 'Coderma' covers three rural parishes to the south of Masaya. The second, 'Proyecto Sol' ('Project Sun') covers several parishes to the north.

People in the two areas had the same problem: they were not on the grid, and although close to Masaya geographically were linked to it only by poor dirt roads, sometimes impassable in the rainy season. Having no electricity meant often having no light at all, or buying expensive (and polluting) kerosene or torch batteries. It also put a lot of limitations on their lives – children finding it difficult to do homework, communities living in darkness (and insecurity) at night-time, and fewer opportunities to develop home-based businesses such as a food shop, carpentry or basket-making.

The majority of the households also continue to use wood for cooking. In the areas south of Masaya, wood is fairly easy to find or buy as there are still extensive woodlands. North of Masaya, however, extensive farming practices in the 1970s wiped out most of the remaining woodland and people often spend a lot of time searching for fuel.[3]

Coderma

Coderma is the name of a co-operative created to run the electricity system for three communities south of Masaya. Their demands to be connected to the electricity grid were finally met in 1999, when a donation from the Japanese government funded a $80,000 project to install the necessary wires, transformers and household connections. The co-operative was formed at the insistence of the donors, apparently against the advice of NRECA, a US-based NGO which promotes electricity co-operatives, who thought the communities were too small and the potential income inadequate. Coderma took over the completed system – which they continued to run when Fenosa took over the national distribution system (providing the co-op with power from the grid at two connecting points).

It proved a disaster. The first co-op committee failed to pay for the power being taken from the national system, despite collecting the money from households. The debt built up until Fenosa took over nationally and cut off the co-operative's supply. Until then, the three communities had no idea that the debt existed. A new committee was formed, and the community agreed to pay again the charges that had never been remitted to Fenosa. But this arrangement too ran into difficulties: the communities became divided around the issue of who should control the co-operative, and debts and maladministration occurred again. For the whole of 2007 the three communities again had no power. Currently (early 2008) power has been restored and a new arrangement about the debt has been reached with Union Fenosa.

Coderma illustrates the pitfalls of benign international aid taking the place of what, in a developed country like Britain, would have been private investment backed up by tight regulation by the state. To get electricity, communities agreed to create a co-operative when they had little experience of running one, and there was no helping hand to guide them through the initial stages of a complex business. A donation that was intended to bring people out of energy poverty and give them energy security only intermittently achieved the first and failed completely to deliver the second aim. An arrangement – creating a co-operative – which was a well-intentioned move to create local control has in fact proved counterproductive.

At the time of writing, the co-operative is negotiating to be taken over by Union Fenosa as the only viable way of controlling what is necessarily a complex and

---

3   The author declares an interest at this point. He lives in the area covered by the first example, and his wife is an activist in trying to solve the problems that have developed.

difficult-to-decentralise service. The co-op members joke that they are the only people in Nicaragua who want Fenosa's services.

Proyecto Sol

Proyecto Sol ('Project Sun') actually originated from my attempt, working with a local Masaya NGO called ADIC, to help with the Coderma impasse – by starting a project to install solar panels as an option for giving people an alternative basic electricity supply. As it happened, it had limited success in the Coderma area, but proved very popular indeed in other parts of Masaya remote from the grid.

The essence of the project is installing 50 watt solar panels, usually on the roof of each house, which feed a 12 volt battery and supply enough power to run 2-3 lights together with a radio or (used sparingly) a TV. The cost is about $1,000 (£500), which is repaid over a period of up to seven years, with no interest. The monthly payment is similar to the price of electricity charged by Fenosa, with the big difference that it pays for an asset which is owned and controlled by the householder.

Compared with Coderma, the capital cost is twice as much per house. But there are virtually no running costs and the system is highly reliable. After being initially sceptical, people are quickly won over by a system which is under their control and whose operation is simple to understand.

Repayments are currently running at about 75 per cent of the optimum level, and this is very high for this kind of project in poor communities. The money is re-invested in new installations, of which so far there are about 65 in total. The capital costs have been funded by donations by UK housing associations, notably Places for People (see Chapter 12) and Midland Heart, with smaller donations from East London HA and from Cairn HA in Scotland.

For 2008, Places for People, LHA-ASRA, Longhurst, Midland Heart and Southern housing groups have already agreed to make contributions that will enable at least another 50 kits to be installed, benefiting over 300 people (and my time as a volunteer continues to be underwritten by my part-time employment with CIH). One attraction of the investment for the associations has been the sustainability of the project – in all senses of the word, but particularly in the sense that each pound invested can be made to work several times over.

The potential for renewable energy sources

It has only been in the last few years that government and lobby groups have started to assess and argue for Nicaragua's capacity to generate its own energy from solar energy and similar renewable sources. There is now ample evidence of what this might be. For example, the United Nations Environment Programme sponsored the international Solar and Wind Energy Resource Assessment (SWERA – http://swera. unep.net/), whose online maps show the tremendous untapped potential in these two (relatively low-investment) fields of renewable energy. In Nicaragua's case, they show considerable solar potential concentrated on the more densely-populated Pacific coast, and wind potential in several regions near to population centres.

Given the country's prior experience with hydro-electricity, it is also not surprising that there have been several major proposals in this field, often controversial. The biggest of all is the Copalar project, which would create a massive artificial lake on the long and broad Rio Grande de Matagalpa, potentially generating enough power to meet all the country's needs and more. Unfortunately it would also wipe out virgin forest, indigenous settlements and invaluable wildlife habitats. Nevertheless, there is considerable potential for smaller, less damaging projects.

Much less controversial are the proven technologies of geothermal energy and biomass, both already operating on a small scale. Also, not surprisingly, biofuels have started to be planted, although the government has set its face against their large-scale development.

An assessment in 2005 by the National Energy Commission suggested potential hydro-electric capacity of up to 1,760 megawatts (MW), geothermal of 1,000 MW, wind of 200 MW and biomass of 100 MW. Even leaving solar power out of this assessment, the potential far exceeds the current installed generating capacity of just over 600 MW.

Interestingly, since its advocacy of electricity privatisation in the 1990s, the World Bank has funded a major solar project in the east of Nicaragua covering several communities, and other NGOs such as Blue Energy (www.blueenergygroup.org/) have become international advocates of solar power, based on their Nicaraguan experience. The obstacle to these systems is cost: $1,000 is out of the reach of most Nicaraguans as a capital investment. But it becomes feasible if it can be repaid over time. Also, support arrangements (like those provided by ADIC) are vital to advise on use of the systems, deal with teething troubles and establish and run a reliable repayment service.

If Nicaragua had only looked across its southern border at its smaller neighbour – Costa Rica – it might have adopted a different strategy at a much earlier stage. In the last 20 years, Costa Rica has promoted renewable energy sources to the point where it is now almost independent of fossil fuels for electricity; and at the same time it has the greatest electricity coverage and the lowest prices in Central America. It also has a publicly-owned system. Nicaragua could clearly achieve what Costa Rica has already done, and end its dependence on fossil fuels to generate electricity – provided it can get the right kind of international help and investment.

Can policy change?

With its experience of developing 'Proyecto Sol', ADIC is now taking part in a project called 'Energy Central,' funded by the European Union under its 'Intelligent Energy' programme, which is aimed at promoting discussion of the wider use of renewables, especially through local and central government. Energy Central brings together various partners in Europe and Central America who are working together to try to shift the climate of opinion, however slightly. One of the partners, IDMEC at the University of Lisbon (a team which specialises in technical and political issues about promotion of renewables, especially in developing countries), has analysed the possible obstacles to progress (see Box 17.1).

Box 17.1: Potential obstacles to the acceptance of renewables

1. Supply isn't continuous – storage (e.g. batteries) needed.

2. High cost of initial investment.

3. Lack of confidence in renewables, either because of ignorance about them or because of resistance to new ideas.

4. Poor or inaccurate information.

5. Perceived risk of being dependent on renewables.

6. Low prices for conventional energy (which don't reflect the 'external' costs such as the environmental damage caused).

7. Poor regulatory environment.

8. Inappropriate political attitudes (e.g. subsidies for conventional energy).

9. Immature market for renewables.

10. Bureaucratic obstacles.

11. Financial obstacles (e.g. can't attract big investors, yet capital costs are high for individuals).

12. Tax system doesn't favour renewables.

13. No effective trade organisation for installers or less effective than the lobby for conventional energy suppliers.

14. Local-level environmental impact (e.g. wind turbines).

15. Distance between sources of energy and main centres of population.

16. Lack of necessary technical skills.

17. Poor image of renewables (e.g. if previous projects have been unsuccessful).

18. Need to refocus energy strategy at national level – a big task.

Source: Presentation by Dr Luis Manuel Alves, IDMEC, University of Lisbon, Portugal.

Most of these obstacles apply in Nicaragua. For example, the Energy Central project hosted meetings with academics, politicians, renewable energy suppliers and community representatives in February and May 2008. Many of the national politicians failed to attend, and the suppliers explained the difficulties they face in trying to change political attitudes and promote a better climate for renewables.

So what does this have to do with environmental issues in Britain?

On the face of it, Britain and Nicaragua have little in common. The UK is ranked 16th on the United Nations' 'human development index', Nicaragua is 110th. In terms of carbon emissions, we Britons collectively churn out 146 times the amount of carbon emitted each year by Nicaraguans.

Yet of course we are interdependent, even if perhaps most people in both countries wouldn't think so. The effects of climate change on Nicaragua have already been mentioned, and are set to get worse as the changes are forecast to affect tropical countries harder than temperate ones. If countries like Nicaragua also start to over-use fossil fuels, and to destroy their forests in the way Europe and North America have done, global climate change will happen even faster.

As we have seen, policies imposed from the developed world have until recently been encouraging countries like Nicaragua to take a similar, fossil-fuel dependant path. As we have also seen, it doesn't work in the short-term and is probably even less sustainable in the future than it is for a country like Britain.

As Simon Retallack of think tank IPPR argued recently:

> 'The single most important issue in the climate change debate is the question of how northern industrialised countries can pay southern industrialising countries to develop cleanly. Unless an answer is found soon, there is no hope of avoiding the worst impacts of climate change.'

He points out that this is a heavy burden, but '...as we have left far too little atmospheric space for the lot of the world's poor to improve along anything like the path the north has, it is a burden we will have to find a way of shouldering'.

But there is a way forward. Many developing countries like Nicaragua could make much greater use of renewable energy sources, if they were helped to do so. Not only would this make more sense environmentally, it could be a more secure and sustainable way of bringing domestic power to the billions of people who lack it.

In the short term, the new Sandinista government that took power in January 2007 has made a mixed start. It secured $100 million finance from the Inter-American

Development Bank to modernise the electricity distribution system and improve the reliability of the main hydro plant. It has reached an agreement (after threats from both sides) to use an empty oil storage facility owned by Exxon for oil imported at favourable prices from Venezuela. The Chavez government in Venezuela has started to build an oil refinery in Nicaragua, and has already installed a small new electricity generating plant. Plans are underway to distribute low-energy light bulbs in poor areas to reduce household bills. Energy-efficient products such as solar panels are now exempt from sales tax. And after challenging rhetoric aimed at Union Fenosa, the government seems to have reached agreement on a way forward to avoid further widespread power cuts.

At the strategic level, the government still urgently needs to develop a plan to provide energy security in every sense of the term mentioned earlier, in which renewables play a major role. It cannot do this alone – it will need outside advice, backed by investment of skills and cash. In the short term, help from Venezuela is invaluable but is hardly reliable, since a change of government on either side might end the relationship.

Yet it is not out of the question to envisage a 10-20 year plan to make Nicaragua much more energy secure, without increasing its carbon emissions, if it received more considered help from the international finance institutions and from other countries than it has received so far. There are some hopeful signs. Nicaragua is to benefit from an $80 million programme to promote renewables launched in 2008 by two of the international financial institutions. And the current energy minister has now set an ambitious (perhaps unrealistic) target of 60 per cent electricity generation from non-hydrocarbon sources, by 2012.

And what does it mean for housing organisations in the UK?

All this may seem an interesting but largely irrelevant issue to housing organisations in the UK. But it isn't. Just as in Britain, there is an enormous battle for political and public acceptance to be won, so that the perspectives and practical ideas sketched out in this chapter can be taken forward. This already happens on a small scale when an individual or organisation offsets their carbon emissions, through one of the many organisations like Climate Care that invest in environmentally friendly projects (such a stoves that consume less wood, or tree-planting schemes) in Nicaragua and other countries. But at present these projects can easily be criticised because of their uncertain impact or because they might have happened anyway.

More costly projects (such as solar installations) may appear to buy fewer carbon savings for each £ spent, but may in the long run be more effective both for the beneficiary and in demonstrating a strategic alternative. There is room for much greater direct engagement by housing associations and others from the UK housing business, not only to achieve change in the UK but to help the right kinds of change to take place in developing countries.

As we have seen, some have done this. Many more associations have some kind of partnership arrangement with a developing country or through a body like Homeless International to help in the global fight against poverty and bad housing conditions. This chapter has tried to show how, just as many associations don't see their social exclusion work as confined to the UK, neither should they see their response to climate change as a UK-only activity. After all, if poverty is a worldwide issue for which the rich north is at least partially responsible, the same applies with even greater force to climate change. The example of Nicaragua perhaps helps to demonstrate this point.

Further reading

This is a list of sources used in compiling this chapter or which provide more general information on the issues raised.

Energy, climate change and poverty in Nicaragua

The monthly magazine *Envio* is available in English and in Spanish (at: www.envio. org.ni) and has regular commentaries on these issues. Particularly useful is: Selma Herrera, Ruth (2005) 'Our Electricity System is One of our Political Class's Great Failures' in *Envio*, October 2005.

Christian Aid has conducted research and campaigns on these issues, as one of the UK NGOs working in Nicaragua (www.christianaid.org.uk). Two reports are particularly useful:
– McGuigan, C. (2007) *The Impact of World Bank and IMF Conditionality: An investigation into Electricity Privatisation in Nicaragua*, Christian Aid
– Christian Aid (2007) *Power and Poverty: World Bank Energy Reforms and Poor People*, Christian Aid Briefing Paper

Energy, climate change and poverty in developing countries

Christian Aid (2006) *The Climate of Poverty*

Jubilee USA Network (2006) *Time for a Clean Energy Revolution*, Briefing paper on oil prices and debt cancellation in poor countries

Retallack, S. (2007) *The Greening of the South* (available at: www.ippr.org/articles/)

Impact of high oil prices and declining supplies on developing countries

Roberts, P. (2005) *The End of Oil: On the edge of a perilous new world*, Bloomsbury (paperback)

See also askquestions.org (www.askquestions.org/articles/oil/oil.pdf).

CHAPTER 18:
Conclusion – the need for action now

Christoph Sinn and John Perry

Much progress has been made since the publication of the earlier CIH book, *Housing and the Environment: A new agenda*, in 1994. The environmental agenda is now part of the political discourse and more importantly has entered the public conscience, and it is safe to say that environmentalism has become part of the mainstream. The crucial significance of climate change has also finally been acknowledged and there is an overwhelming consensus amongst the scientific community that man-made greenhouse gases are largely to blame for global warming. In the words of Achim Steiner, executive director of the UN Environment Programme, the fourth annual report by the Intergovernmental Panel on Climate Change (IPPC) in 2007 effectively puts an end to the debate about whether human activity is responsible for global warming. The report puts its likelihood at 90-95 per cent, or, as the IPPC puts it, *'very likely'* (IPCC, 2007).

However, in spite of all the available evidence (and unlike many other current geo-political issues), John Lanchester (Chapter 2) shows that climate change is still bitterly contested. This makes it all the more difficult to reach a consensus in terms of a plan of action that really addresses the problem. However, Lanchester is in no doubt (and the editors and authors of this book would agree) that the issue demands urgent action on a scale yet unseen; a global response to a truly global crisis.

Most of the suggested actions in this book require not only strong political will but even stronger political leadership and the adoption of a long-term perspective beyond the usual five-year election cycle. They also call for an enlightened population, prepared to forego many of the things taken for granted for such a long time. This book has the modest aim of making a positive contribution to this cultural change within the housing sector, by bringing the issue to a wider, non-specialist audience.

So far in the UK, we fall far short of taking (or even agreeing) decisive action. Many will argue that even if the UK was to adopt the most radical carbon reduction measures, it would not make a difference globally, mainly due to the energy intensity of China's and India's rising economies. In fact, only recently China has overtaken the US as the world's worst polluter (Vidal and Adam, 2007). We don't disagree with such reservations for a single moment, but would argue that if the UK is to play a leading role in combating global warming on the international stage (for instance in securing a comprehensive post-Kyoto settlement) then it has first to get its own house in order.

It should also be borne in mind that per capita emissions in the UK are significantly higher than in China and India (Table 18.1).

Table 18.1: Residential CO_2 emissions per capita in 2003 (kg per person)

UK	1,412.6
China	185.6
India	90.1

Source: Earth Trends/World Resources Institute.

The principle of leading by example is an important theme throughout the book. The Climate Change Bill and the zero-carbon targets for new build are – it could be argued – the first steps in establishing such a leadership role, setting a precedent internationally. However, we should be under no illusion that statutory targets on their own are sufficient to bring about the desired change, as Peter Lehman (Chapter 6) points out. In the context of fuel poverty he argues that the existence of a statutory target has made a real difference, insofar as it has helped secure more resources for improving the homes of vulnerable people and greater engagement across government. But it has not made *enough* difference. A great deal of additional work will be needed to put in place the actual measures required to meet these targets.

Whilst these initiatives are steps in the right direction, there are a many contradictions at the policy level. For instance, emissions from shipping and aviation are not part of the proposed climate legislation. Recent studies put global emissions from shipping at four per cent, double that of aviation. With global trade set to increase, it is estimated that emissions from the global shipping fleet will rise by 75 per cent over the next 15 years (Vidal, 2007). Since Britain relies heavily on the shipping industry for imports and exports, emissions from this source need surely to be taken into account when setting emission caps.

The same should apply to aviation, which also does not feature in the government's emission reduction targets, although it is common knowledge that the warming properties of greenhouse gases emitted by planes are much greater than those from other fossil fuel-burning sources. It simply does not square up to propose a massive airport expansion programme and at the same time draw up legislation which is meant to pave the way for a low-carbon and essentially post-oil economy.

Many would argue that stemming global warming is a fundamental challenge to the status quo of current economic practice. But given the recent crisis of financial markets which have been left in disarray and a looming global recession, questioning the prevalent economic paradigm might not fall on deaf ears after all. It is simply not possible to 'save' the planet and pursue our goal of rampant economic growth. Yet for many, the 'holy grail' is to resolve the dichotomy of continuing growth in the world economy while at the same time saving the planet.

However, the fallacy of this thinking had to some degree been exposed some time ago with the publication of the *Limits to Growth* report by the Club of Rome, an informal group of scientists, researchers and industrialists (Meadows *et al.*, 1972). The group focused on what it saw as the five issues of global concern, namely accelerating industrialisation, rapid population growth, widespread malnutrition, depletion of non-renewable resources, and a deteriorating environment. Their thesis was based on a series of computer models, simulating future states of the world by altering set variables (i.e. industrial output per capita, population, food per capita, resources and pollution). All models pointed towards the same scenario: that the limits to growth are eventually reached, regardless of the measures adopted. Whilst the methodology was riddled with flaws, and most of its conclusions grossly overstated and didn't come to fruition (Connelly and Smith, 2003), *Limits to Growth* was nevertheless very important in challenging assumptions about humanity's unconstrained use of finite resources in a finite system. That this line of reasoning is basically correct is confirmed in the United Nations' recent Millennium Ecosystem Assessment (see Chapter 3).

When it comes to climate change, the crucial difference is that there is much more *certainty* about the potential impacts of non-action. The *Limits to Growth* thesis (for all its flaws) also reminds us of an important principle which underpins environmental policy and politics: the 'precautionary principle'. It states that in the light of the uncertainty as to the potential (negative) impacts of certain activities and technologies on the environment, we should adopt a position which assumes that these threats are real and thus stop or at least limit such activities.

Even if those negative impacts prove to be unfounded, we won't be worse of, since the measures put in place to compensate for certain activities will have created new opportunities. In the case of climate change this means putting in place measures to drastically cut our CO_2 footprint. Should science be proved wrong about human-induced climate change, our diversionary strategies will have given rise to new economic opportunities and helped to create healthier and cleaner environments. We admit that this seems rather an over-simplification, but when considering the housing issues with which this book is concerned, then it won't appear that far off the mark.

As the book has shown, investment in a low/zero-carbon housing will have numerous benefits, not just for the environment. A 'green' home creates a healthier indoor environment, uses less energy, produces less waste, which in turn means less energy and resource wastage, lower fuel bills and better health for occupants and a reduction in the number of fuel-poor households. On a wider scale it influences travel and transport patterns, instils 'pride of place', can lead to more environmental awareness and addresses the issue of fuel security and helps to create new jobs – to name only a few of the advantages. As evidence for the latter, one only has to consider the case of Germany. It is estimated that for every £1 billion invested in the existing housing stock, around 25,000 new jobs are created (Jung, 2007).

Whilst we do not see ourselves as adherents to a 'deep-green'/ecological school of thinking,[1] which often underlies the debates about the reconciliation of environmental and economic goals, we nevertheless believe that changes to the way we currently live are inevitable if we are to succeed in averting the worst-case climate scenario. In the context of housing, the well-accepted notion of sustainable development is a useful way of looking at this question, since it has been shown to be a '...*powerful tool for consensus*' (Repetto, 1986, cited in Lélé, 1991). As such it can help to find a 'mid-way' point between the two ends of the 'green' political spectrum.[2]

The most widely known definition of sustainable development is that formulated by the Brundtland commission, as '...*development that meets the needs of the present without compromising the ability of future generations to meet their own needs*' (World Commission on Environment and Development, 1987). Although it has been repeatedly criticised for being '*sufficiently vague*' (Garner, 2000), the core principles of the concept are as relevant to housing today as they were at the time of the earlier book, *Housing and the Environment: A new agenda* (Bhatti et al., 1994).

First, there is the principle of inter-generational equity (also sometimes called 'futurity'). This means that all our present activities have to be conducted in such a way that future generations are still able to fulfil their own needs and aspirations. Making the UK's existing housing stock fit for the 21st century, in terms of adapting it to a changing climate but perhaps more importantly unlocking the huge carbon-saving potential it presents, is a key way of applying this first principle. After all, over 80 per cent of existing homes will still be in use by 2050 (Boardman, 2007). Equally, as referred to earlier, there are huge employment opportunities in the construction and renewable technology sector.

---

1   A good starting point to shed light on the 'deep-green' school of thought is Dobson's *Green political thought*, and Connelly and Smith's *Politics and the Environment*.

2   These can be explained using Dobson's (2000) distinction between 'environmentalism' and 'ecologism'. He says that the former suggests a '...*managerial approach to environmental problems, secure that they can be solved without fundamental changes in present values or patterns of consumption*', whereas the latter argues that a '...*sustainable and fulfilling experience presupposes radical changes in our relationship with the non-human natural world and in our mode of social and political life*'.

Second, at the heart of sustainable development lies concern for the global environment, which is being jeopardised due to human activity. It needs to be acknowledged that the principle of ecological sustainability has been somewhat 'watered' down in the years following the Brundtland report, as it now tends to incorporate policy areas such as health, education and social welfare (Baker *et al.*, 1997). This is not a problem as such, since it has been recognised that environmental and social concerns are inextricably linked (Tuxworth and Porritt, 2002). But it does become a problem if environmental concerns are displaced by these or other objectives.

This is an important point, as it might be argued that the popularity of sustainable development as a concept is down to the fact that it doesn't necessarily challenge current political and socio-economic realities. In fact, sustainable development *promises* that environmental sustainability and economic growth can be achieved at the same time (Lélé', 1991). However, we would argue that continuing economic growth is not necessarily an issue in itself, providing it *supports* and *drives* environmental and social objectives and is therefore genuinely sustainable.

The link between the environmental and the social is perhaps most evident when considering the unequal distribution of so-called environmental 'bads' (Cutter, 1995). The point has been made that poorer sections of society are more likely to live in areas with high traffic densities, pollution and badly insulated and poorly maintained housing (Tuxworth and Porritt, 2002). In short, such areas are characterised by environmental blight and degradation. This argument, incidentally, is stronger still if applied to the slums in cities in the developing world (as demonstrated in Mike Davis's *Planet of Slums*).

Research in the early part of this decade by Friends of the Earth for instance has shown that there is a clear link between the socio-economic profile of a neighbourhood and the distribution of air pollution from factories (Friends of the Earth, 2001). By comparing the location and the levels of emissions from local factories with the government's indices of deprivation, it has been shown that the most deprived wards are also the worst polluted.

The unequal distribution of negative environmental impacts is of course not only confined to pollution from industries. Health inequalities also have a clear environmental dimension to them. It might be assumed that the health impacts of car pollution for instance are shared among all city dwellers, but this is not the case. It has been shown that respiratory problems in London are concentrated in the poorest areas and are linked to high traffic levels, yet these areas have the lowest levels of car ownership (Stevenson *et al.*, 1991, cited in SDC, 2002). Although the link between socio-economic status and environmental inequalities has only fairly recently been acknowledged in the UK, it became the focus of a whole new movement under the banner of environmental justice in the US (Agyeman, 2000).

Housing policy (and also spatial planning – see Chapter 5), can play a crucial role in addressing these issues. This is already evident in the policy emphasis on 'mixed tenure' estates and communities,[3] an area in which social landlords can take the lead (see for example Chapter 12). Equally, local authorities with their 'place shaping' role and 'wellbeing' powers (see Chapter 4), are key to addressing socio-spatial inequalities.

The third principle is that of intra-generational equity or social justice. Environmental degradation/climate change threatens the livelihood of millions of people, even in our generations. It is well known that the poorest communities will be (and in many instances are already, as the preceding chapter has brought home) worst hit by a changing climate. The activist and writer George Monbiot put it rather eloquently when he said that '...*shoddy building work in Exeter kills people in Ethiopia*', as a way to highlight the global impact of what might at first seem to be a rather minor issue (Monbiot, 2006). Research by the Energy Efficiency Partnership for Homes amongst Building Control Officers on compliance with Part L1 of the 2002 building regulations (conservation of fuel and power) revealed that on the whole officers will not enforce or refuse completion certificates or prosecute on a Part L issue (EEPH, 2006). Part L was given low priority as it was seen as being not *life threatening*. This practice clearly has to change and in order to achieve zero-carbon, energy-efficiency requirements need to be enforced rigorously. The fact that, since the introduction of energy conservation legislation 21 years ago, not a single builder has been prosecuted for non-compliance is a reminder that environmental awareness has yet to become mainstream thinking within the building industry (Monbiot, 2006).

Furthermore, unequal access to resources has a detrimental effect on the environment, particularly in the 'majority world' (but also closer to home, as in the case of fuel poverty – Chapter 6). John Perry in Chapter 17 demonstrated how the 'well-intentioned' policy of privatising electricity distribution led to a volatile and inefficient system, with the poorest affected worst, giving rise to widespread 'energy poverty'.

The final principle is that of greater community participation. In order for sustainable development to succeed, it is necessary that communities are able to have a say prior to and when decisions affecting them are being taken. Whilst more and more people accept the science of climate change, getting them to adopt more environmentally friendly behaviour proves to be more challenging. For instance, as Chapter 10 has shown, there are just about 100,000 homes using low- and zero-carbon energy technologies, which amounts to about one micro-generation measure for every 300 homes.

---

3   There is quite a mature body of literature around this and allied issues and readers might be interested in a number of recent publications, produced by the CIH for the Joseph Rowntree Foundation (Allen *et al.*, 2005, Rowlands, *et al.*, 2006, Silverman *et al.*, 2006, Bailey *et al.*, 2006).

Social landlords are in an excellent position to engage with residents and the wider community around environmental issues. For instance, research examining local sustainable development projects in deprived neighbourhoods found that apart from their social and economic benefits (e.g. developing new skills, creating employment opportunities), such initiatives are an important vehicle for raising environmental awareness and prompting action in communities, which otherwise would not have prioritised these issues (Church and Elster, 2002). In other words, tackling 'local' environmental issues is a first step towards global sustainability. The local environment is where the well-known slogan *think globally, act locally* should come to fruition.

The book has made it clear that housing is pivotal to meeting the government's target to reduce carbon emissions by a minimum of 60 per cent by 2050. The carbon reduction potential of the housing sector is huge, as is the potential for cost-effective 'quick-win' solutions. Whilst the current government emphasis on increasing the environmental performance of new build is laudable, the government is in danger of neglecting the biggest problem facing us – emissions from existing homes.

We believe the time is right for a massive structured programme which deals with the existing stock in a strategic and comprehensive manner, with investment at levels along similar lines to past inner-city renewal programmes. The current 'piecemeal' approach does not address the scale of the problem and is not fit for purpose. What we don't need are yet more pilot schemes and initiatives. The book's most important conclusion – its 'bottom line' – is that without investment in a wholesale transformation of UK housing, which includes the existing as well as newly built stock, the government will completely fail to meet its carbon reduction targets.

References and further reading

Agyeman, J. (2000) 'Environmental justice – from the margin to the mainstream?', *Town & Country Planning*, Vol. 69, No. 11, November, pp. 320-322

Allen, C., Camina, M., Casey, R., Coward, S. and Wood, M. (2005) *Mixed tenure twenty years on: Nothing out of the ordinary*, Coventry: Chartered Institute of Housing

Bailey, N., Haworth, A., Manzi, T., Paranagamage, P. and Roberts, M. (2006) *Creating and sustaining mixed income communities: A good practice guide*, Coventry: Chartered Institute of Housing in association with the Joseph Rowntree Foundation

Baker, S., Kousis, M., Richardson, D. and Young, S. (1997) 'Introduction: the theory and practice of sustainable development in EU perspective', in S. Baker, M. Kousis, D. Richardson and S. Young (Eds.) *The Politics of sustainable development: theory, policy and practice within the European Union*, London: Routledge

Bhatti, M., Brooke, J. and Gibson, M. (1994) *Housing and the Environment: A new agenda*, Coventry: Chartered Institute of Housing

Boardman, B. (2007) *Home Truths: A low-carbon strategy to reduce UK housing emissions by 80% in 2050*, Environmental Change Institute, Oxford University. Available from: http://www.eci.ox.ac.uk/research/energy/hometruths.php [Accessed10 December 2007]

Church, C. and Elster, J. (2002) *Lessons from local action for national policy on sustainable development*, Joseph Rowntree Foundation. Available from: http://www.jrf.org.uk/knowledge/findings/housing/522.asp [Accessed 22 February 2003]

Connelly, J. and Smith, G. (2002) *Politics and the Environment: From Theory to Practice (2nd edition)*, London: Routledge

Cutter, S. L. (1995) 'Race, class and environmental justice', *Progress in Human Geography*, Vol. 19, pp 111-122

Davis, M. (2007) *Planet of Slums*, London: Verso Books

Dobson, A. (2000) *Green political thought (3rd edition)*, London: Routledge

Earth Trends (World Resources Institute). Available from: http://earthtrends.wri.org/index.php [Accessed 13 April 2008]

EEPH (2006) *Compliance with Part L1 of the 2002 Building Regulations: an investigation on the reasons for poor compliance*, Energy Efficiency Partnership for Homes. Available from: http://www.eeph.org.uk/resource/partnership/index.cfm?mode=view&category_id=35 [Accessed 17 February 2007]

Friends of the Earth (2001) *Pollution and Poverty – Breaking the link*, Friends of the Earth. Available from: http://www.foe.co.uk/resource/briefings/pollution_and_poverty.pdf [Accessed 10 February 2003]

Garner, R. (2000) *Environmental Politics (2nd edition)*, London: Macmillan

IPCC (2007) *Climate Change 2007: Summary for Policymakers*, Intergovernmental Panel on Climate Change. Available from: http://www.ipcc.ch/pdf/assessment-report/ar4/syr/ar4_syr_spm.pdf [Accessed 25 May 2007]

Jung, A. (2007) 'Die beste Energie: Sparen', *Der Spiegel*, 12 February

Lélé, S. M. (1991) 'Sustainable development: A Critical Review', *World Development*, Vol. 19, No. 6, pp. 607-621

Meadows, D. H, Meadows, D. L, Randers, J. and Behrens III, W. W. (1972) *The Limits to Growth: A report for the Club of Rome's Project on the Predicament of Mankind*, New York: Universe Books

Monbiot, G. (2006) 'Strange but true: shoddy building work in Exeter kills people in Ethiopia', *The Guardian*, 30 May. Available from: http://environment.guardian.co.uk/climatechange/story/0,,1828830,00.html [Accessed 17 February 2007]

Rowlands, R., Murie, A. and Tice, A. (2006) *More than tenure mix: Developer and purchaser attitudes to new housing estates*, Coventry: Chartered Institute of Housing

Silverman, E., Lupton, R. and Fenton, A. (2006) *A good place for children? Attracting and retaining families in inner urban mixed income communities*, Coventry: Chartered Institute of Housing

SDC (2002) *Vision for sustainable regeneration: environment and poverty – the missing link?*, Sustainable Development Commission. Available from: http://www.sd-commission.org.uk/publications/downloads/021001%20Vision%20for%20sustainabl e%20regeneration,%20environment.pdf [Accessed 20 October 2002]

Tuxworth, B. and Porritt, J. (2002) 'Regeneration: It starts with an E', *Green Futures*, May/June, pp. i-ii

Vidal, J. (2007) 'CO$_2$ output from shipping twice as much as aviation', *The Guardian*, 3 March. Available from: http://environment.guardian.co.uk/climatechange/story/ 0,,2025726,00.html [Accessed 10 March 2007]

Vidal, J. and Adam, D. (2007) 'China overtakes US as world's biggest CO$_2$ emitter', *The Guardian*, 19 June. Available from: http://www.guardian.co.uk/environment/ 2007/jun/19/china.usnews [Accessed 20 June 2007]

World Commission on Environment and Development (1987) *Our Common Future*, Oxford: Oxford University Press

More books from CIH and HSA

INTRODUCING SOCIAL HOUSING

Stephen Harriott and Lesley Matthews

The provision and management of social housing demands a wide range of knowledge and expertise. This important book provides a stimulating introduction for both students and staff needing to understand *what social housing is* and *how it can be most effectively provided*.

'Partnering' approaches to social housing provision are very high profile, therefore it is not only housing staff who need to be aware of recent and anticipated changes in housing finance, development, management and the law. Planners, social workers, developers and many other professionals need to understand what social housing is all about.

Introducing Social Housing offers a readable description and thorough analysis of:

- how social housing has evolved;
- the range of tenures;
- the nature of running a 'social business' and achieving Best Value;
- the social context of changing priorities for lettings and meeting the needs of different tenants;
- core housing management functions and skills such as rent setting and collection, maintenance and dealing with anti-social behaviour;
- achieving customer involvement and satisfaction;
- dealing with housing supply and demand questions, and
- how the provision of social housing forms an important element in regeneration, what the government calls 'liveability' and future economic growth.

This book is ideal for those studying for the Chartered Institute of Housing recognised qualifications, and also for other social policy courses where an understanding of social housing policy and practice is required. But it will also be invaluable for professionals in housing and other fields who need to have a good understanding of the background to current policy.

This book has been completely revised and the new edition will be available in September 2008.

Publication of the new edition of *Introducing Social Housing* has been sponsored by the Aster Housing Group.

Order no: 128
ISBN: 978-1-905018-63-5
Price: £25.00
Available: September 2008

SAFER COMMUNITIES – HOUSING, CRIME AND NEIGHBOURHOODS

Edited by Alan Dearling, Tim Newburn and Peter Somerville

The challenge is to make communities safer, and this book is about the key issues which those working in and with communities now face in trying to meet this challenge. It argues that a balanced and proportionate response is needed to anti-social behaviour and crime – one that offers strategies for prevention, intervention, support and enforcement.

Fifteen thought-provoking chapters address the themes of 'disorder and regeneration', the 'policing of crime and disorder', 'service provider approaches to safe communities' and 'social inclusion and community safety.'

The authors explore the importance of working with residents in community safety responses and examine how staff in housing, social work, police and education can overcome barriers by working in partnership. They recognise also that we need robust housing management practices that respond to tenants' support needs.

Housing professionals will find examples of effective strategies for dealing with the wide range of anti-social behaviour and criminality that can exist in communities today. They will also have access to the policy debate about how we tackle crime and fear of crime, and the relevance of and the questions that need to be asked about the government's 'Respect' agenda. The book is also highly relevant to all those working with community groups or in neighbourhood regeneration, in areas where crime, fear of crime or anti-social behaviour are on the agenda.

Publication of *Supporting Safer Communities* has been sponsored by Affinity Sutton Group.

Order no: 125
ISBN: 978-1-905018-30-7
Price: £25.00

HOUSING FINANCE

David Garnett and John Perry

A completely revised edition of the definitive guide that has proved to be one of the best introductions to housing finance across the UK.

This edition deals comprehensively with all housing sectors and provides both a policy context and a wealth of up-to-date information on each topic. The book is a perceptive analysis of the present system of housing finance as well as a detailed review of all recent changes. It deals with developments such as the prudential borrowing regime, the investment partnering arrangements for housing associations, the latest incentives to home ownership, the rapid growth of arms length management organisations, rent restructuring – and parallel developments in Scotland, Wales and Northern Ireland. It also deals with business planning and treasury management.

This is the first book to provide such an overview since the overhaul of most aspects of housing finance that has taken place over the last few years.

The 500-page guide sets the conceptual context, then deals with the role of the state, has eleven chapters covering capital and revenue finance in each of the four main housing tenures, and a chapter describing housing benefit and other support for housing costs.

Publication of *Housing Finance* is sponsored by the Aster Housing Group.

Order no: 121
ISBN: 1-903208-53-X
Price: £27.00